Consultant Role in Construction Projects- Electrical Installations

Consultant Role in Construction Projects-Electrical Installations

Author,

Nedal Nabulsi

Co author,

Hani AlKhayyat

WINGS PUBLICATION INTERNATIONAL

WE GIVE WINGS TO YOUR DREAM!

Disclaimer

WINGS PUBLICATION INTERNATIONAL
WE GIVE WINGS TO YOUR DREAM!

Registered Office- 907-Sneh Nagar, Sapna Sangeeta Road,
Agrasen Square, Indore – 452001 (M.P.), India
Website: http://www.wingspublication.com
Email: mybook@wingspublication.com

First Published by WINGS PUBLICATION 2024

Title : Consultant Role in Construction Projects- Electrical Installations
Price : AED 150 | $ 41

ISBN : 978-93-6006-736-6

LIMITS OF LIABILITY/DISCLAIMER OF WARRANTY

Dedication

Dedicated to:
All our teachers and lecturers
All our seniors
All our colleagues
All fresh engineers who would like to gain experience.
All people who supported and participated in this book directly or indirectly.

Acknowledgement

This book is a result of continuous work in the construction field that led to a massive experience that was collected by cooperation with many experts, engineers, and seniors.

Hence, special thanks to all the people who were a source of information by teaching, directing, guiding, and advising.

Special thanks to our teachers and lecturers throughout our study life.

Thanks to all colleagues and staff who worked with us on the construction projects and participated in collecting, providing, submitting, and sharing the information or printing it.

Special thanks to Wings publication and Dr. Kailash Pinjani for his great support, encouragement, and push to make this work appear to the readers.

Legend

Preface

This book is the accumulation of massive experience in construction project consultation for the installation, selection and coordination of electrical systems and work.

The authors found that the entry-level practical information required to start the electrical consultant duties on the site is not always available. Several fresh graduates asked to be taught about the consultant role in the project and the duties he need to do to be a key member in the work and to be able to continue till completion and to continue in the next projects.

While such expertise is mainly practical expertise that can't be offered in academic colleges, the deficiency in such expertise may lead to huge consequences in the project life, especially if it leads to wrong installations or deviations from the standard, regulations or client requirements that can affect the project function or reduce the installation value.

In addition, fresh graduate engineers usually work with seniors, so they don't have enough time to be trained properly due to the huge load on the staff in the construction stage. This leads to a lengthy period of collecting experience from others and will require a lot of work in the entry-level site work.

Hence, this book is prepared for fresh engineers who would like to gain experience in a short duration by reviewing a massive collection of practical site experiences that are offered in the book.

We can promise that the information shown in this book will give the reader at least 5 years of experience that can be gained just by reading the book and understanding its content.

The book was started with some definitions and abbreviations that may be required by newly graduated engineers and divided into several chapters to explain the following:

1. The required work in the starting stage of the project.
2. The required drawings and the importance of each.
3. The required coordination with other systems, especially the mechanical systems. That contain most of the huge loads.
4. The required containment for the wiring and the cables.

5. The installations of the electrical systems are divided into 6 chapters.
6. The conclusion.

The book was divided system-wise. Hence, the reader should be aware of the priority of work and the priority of the activities and systems he needs to complete on the site to progress smoothly with the main contractor work and other systems installations.

The chapters were divided to explain every system requirement for correct and proper installation in a dedicated chapter. Many of the points mentioned in each system are repeated, and that was done intentionally to allow smooth reading without forcing the reader to jump from one page to another and loose reading continuity.

Finally, we would like to thank all people behind this achievement and would appreciate your feedback to improve the book in the next revisions.

The Authors.

Nidal Nabulsi

Hani AlKhayyat

Abbreviations

A/C	Air conditioning system
ACB	Air circuit breaker
AHU	Air handling unit
ARM	Armored
ATS	Automatic Transfer Switch
AWA	Aluminum wire armored.
BMS	Building management system.
BOQ	Bill of quantities
BTU	British thermal unit
CB	Capacitor Bank
CBU	Central Battery unit
CCTV	Close circuit TV system
CCU	Close control unit
CP	Control Panel
CU	Condenser unit
DB	Distribution board
DDC	Direct digital control
DOL	Direct online
EBD	Emergency battery drive (lifts)
EBS	Emergency Battery system
ECC	Earth copper cable
ELCB	Earth leakage circuit breaker.
ELR	earth leakage relay
F/C	false ceiling
FACP	Fire Alarm control panel
FAHU	Fresh air handling unit
FCU	Fan coil unit
FDP	Feeder Pilar
FFL	finish floor level
FP	feeder panel
GI conduits	Galvanized Conduits
GIT	Galvanized Sheet
GRMS	Guest Room Management System
GSM	Global System for Mobile Communication.
HT	High tension cables
HV	High voltage
IC	Intercom
IDF	Intermediate distribution frame
IF	Intermediate frequency

IMS	Intelligent Metering system
IP	Ingress protection
IR	Inspection request
JB	Junction Box
LFP	Lighting feeder panel
LSZF	Low smoke zero fume
LSZH	Low smoke zero halogen
LV	Low Voltage
M/C	Machine cleaning room
MATV	Master Antenna TV systems
MCB	Miniature circuit breaker
MCC	Motor Control Center.
MCCB	moulded case circuit breaker
MD	main distribution
MDF	Main distribution frame
MDU	main distribution unit
MEP	Mechanical Electrical Plumbing
MGT	Mica glass tape
MICC	Mineral insulated copper cable.
MIR	Material inspection request.
NCS	Nurse calling system.
NOC	No objection certificate
PA	Public Address System
PBX	Privat Branch Exchange
PC rate	Primary cost rates
PMS	Power management system
Psum	Provisional sum
PS	Power supply
PVC	Polyvinyl chloride
RCC	reinforced concrete columns.
RCCB	Residual current circuit breaker
RF	Radio frequency
RFI	Request for information
RMS	Radio messaging system
RMU	Ring main unit.
RYB	Red, Yellow, Blue
SEF	Smoke exhaust fan
SL and DC	Smart Lighting and Dimming Control System
SS	Soft starter
SWA	Steel wire armoured.
TEL	Telephone
TV	Television
UPS	Uninterruptible power supply
VFD	variable frequency drive
W/H	Water heater
W/P	weatherproof
XLPE	cross-linked polyethene

Definitions

Activity chart or program: A chart that shows the sequential activities to be done on the site, with dates, expected duration and dependencies to other activities.

Addendums: Any additions to the original scope of the contract or a specific document.

Attic Slab: A slab is to be added over the electrical panel to prevent water flow from above the floor if it contains a bathroom, toilet or kitchen; the Slab should have an opening that allows the maintenance team to climb to reach the above floor services (the slab can be metallic).

Authority Approval (NOC): No objection certificate is being granted if the application fulfils the authority requirements.

Authority regulations: The rules and requirements that are mandatory in the work based on the authority recommendations, usually published and updated continuously.

Back of house: The areas where the installation is far from the eye site, like over false ceiling or in service rooms.

Builders' work drawings: these are the drawings related to the built walls and slabs that describe the required openings for MEP work.

Cause and effect matrix: These are the inputs and outputs for a system that shows intiating and output signals.

Compliance statement: This is the procedure for the supplier to show how accurately he complies with the specifications; it is done by printing the original stamped specification and marking each paragraph as complying. If not complied, then he should show the deviations clearly to enable the consultant to decide if it can be accepted or not.

Containment: These are the materials used to contain the wires and cables like conduits, trunking, trays, and baskets.

Coordination drawings: These are the drawings showing all systems and equipment for all trades in the area with accurate dimensions and all dimensions between them. The coordination drawings will show the discrepancies and the clashes between systems and the wrong distribution due to the effect of other systems.

Cost and time implications: When there is a variation in the project, then the contractor will agree to implement the work based on the approval to extend the time of the project and increase the cost for the new work.

Cost saving: When there is a change in the project work that reduces some work or changes the system to a cheaper one, then the consultant would agree (if all specifications can be met) based on the reduction of the paid money (cost saving).

Design drawings: The design drawings are the set received from the designer; they include information about the required systems only, without any details of the size, dimensions, containment, peripherals, and the effect of systems on each other.

Environmental conditions: These are the project environments that the panels and the equipment will suffer in their lifespan, like the temperature, humidity, altitude, and how close to sea level they will be. All supplied systems should be able to work within the project's environmental conditions.

False ceiling clearance: The clearance is a confirmation from the MEP subcontractor that he finished the installation of all services over the false ceiling, and the main contractor can close it. Hence, if anything is missed in the future, re-opening will cost the MEP subcontractor the cost of opening.

Front of house: This is the most luxurious location in the project, including entrances, lobbies, residential flats, and everything other than the service rooms.

Good for construction drawings: These are the drawings received from the designer, which may include some additions to the design drawings and may include some missing systems, but it is still enough to start the construction till receiving the rest of the design.

Ingress protection: This is based on the standard IEC60529 for the requirement to protect the components or equipment from water and dust, and it is marked by IP(XX), where XX can be from 0 to 6 for both digits).

Input-output matrix: See the cause-and-effect matrix.

Inspection request (IR): an official request through the special template to be sent to the consultant to inspect the work or material to get official approval from the consultant or to show his comments that need to be rectified.

Interfacing systems: These are the related systems that will function a certain action triggered by the main system, ex lifts will stop in the nearest floor to evacuate passengers and then travel to the ground floor, this will occur when fire alarm is initiated.

Interlocking systems: Systems that, when become on, they switch off or cut signals for other systems.

Item breakdown: Dividing a task into several parts to control work and pricing, ex the generator installation task can be broken down into fixing the generator, cable laying and termination, fixing the fuel tank and its pipes.

Maintained Emergency lighting: These are the emergency lights that are always on due to connection to power supply. When the power supply is cut, the battery will feed the power and will keep it on like exit lights.

Markup drawings: These are copies of the approved shop drawings with markings for the installation deviations in the site that have been done due to site requirements, site conditions or variations. They are used to generate the as-built drawings.

Material inspection request: Inspection request for material approval by the consultant (see inspection request).

Method statement: this statement is submitted by the contractor showing in complete detail the way he is going to do an activity by showing:

1. Brief description of the project.
2. Purpose.
3. Scope of work
4. Responsibilities of persons or contractors.
5. Used tools.
6. Safety requirement.
7. Details and description of work.
8. Coordination with others.
9. Attachments (drawings, tables, etc.).

Mobilization: It is the movement of the contractor team to start the work on the site. Usually, they start with the site office installation for the working team of the contractor and the consultant.

Mockup: It is a trial installation for an activity to be approved by the consultant for implementation in the site, like a mockup of wall painting or a mockup of the façade installation.

Mounting height schedule: The schedule contains the height of all electrical accessories (switches, sockets, thermostats, bells, sounders, etc.). The mounting height schedule should be approved by the consultant to be implemented on the site; it should be updated and resubmitted when additional accessories are to be installed.

Nonconformance report (NCR): This report is initiated by the consultant when deviation from the approved drawing, approved material, and approved installation is found on the site, which affects work quality.

Non-maintained Emergency lighting: These are the emergency lighting that is in an **off** state by connection to the power supply. When the power supply is cut, the emergency lighting battery will feed the power and switch it on.

Override switches: These switches near the lighting control panel will override the lighting control program and will switch the lighting permanently.

PC Rates (Primary cost rate): This is the cost of material with an unknown price. The BOQ will contain the assumed cost. The contractor will submit the material that will be used in the site, If approved, the actual cost of the used material will be compared with the assumed cost in the BOQ, and based on the result, positive variation to the contractor will be approved (if the used material is more costly than the BOQ price), and a negative variation will be created if the actual cost is less than the BOQ cost.

Provisional sum: This material will be procured by the project owner or the client but will be installed by the contractor; the contractor will be paid the installation fees only.

Request for information (RFI): Official letter to be sent to the consultant, client, or designer for clarification of some items when there is missing information or discrepancies in the design.

Ring main unit (RMU): A switching unit used in the high-tension network to allow tapping in the high-tension network to the load.

Site instruction (SI): An official instruction within a special form on the site used by the consultant when he finds any point or instruction that requires to be noticed by the contractor.

Snag or Punch list: At the end of the project, the consultant and the contractor will check the installed systems and will check that all inspection requests' comments are closed and all installations were rectified according to the consultant's comments. These comments will be added to the punch list to be reviewed after an agreed duration.

Stakeholders: All parties related to the work, whether contractor, consultant, suppliers, client, authority or any other party, can be affected or related to the work.

Tender drawings: These are the drawings to be given to the contractors for tendering and bidding. These drawings should be compared with the drawings received on the site to know the additions to the tender.

Testing certificates: The testing done on the received material by official and reputed testing companies.

Trunking space factor: The amount of trunking cross-section area to be filled by the wiring; usually, it is 45% to allow proper ventilation for the wiring.

Variation: Any additional requirement by the client to the scope of work will be a variation in the project. The variation should be approved by the client and the consultant for price and required duration.

Water table or water level: It is the level of the underground water that will appear when excavating the area.

Workshop drawing: these are the drawings prepared by the contractor that includes:

The final location of the equipment.

The used equipment with actual dimensions.

The dimensions between equipment.

The details of containment and wiring.

The details of installations.

The final load based on the approved material ratings and approved mechanical loads.

The final interfacing with all systems.

Shop drawing should be approved by the consultant.

Chapter 1

Preparations before Starting the project

Points To Be Studied and Notified to The Contractor Before Starting Construction:

Before starting and in the mobilization stage, the consultant should follow and instruct the contractor to study the following and notify the consultant and the client of any discovered discrepancies:

1. **Drawing's comparison.**

The consultant should follow the contractor to ensure that there are no differences or discrepancies between design drawings/ good for construction drawings handed over to the contractor and the original approved design drawings granted the No Objection Certificate (NOC) by the Authorities (Power provider, telephone service provider, civil defence...etc.).

2. **Design Verification.**

The consultant should instruct and follow the contractor to verify the design and to check if some systems can be installed in different ways that may reduce the cost and working time or increase the quality while still following the applicable standards and the authority regulations. Such modifications may require re-calculations and may cause cost savings, for example:

   a. Lighting in the basement on raceways is to be changed to ceiling-mounted lighting with PVC conduits in the slab (may require lux re-calculations due to raising the fittings up).

   b. Power cables in conduits with manholes are to be changed to buried cables in the ground.

   c. Changing the location of any panel may require drop voltage re-calculations and may lead to cable size changes due to increased length or installation method changes.

3. **Provisional Sum and PC rate.**

The consultant should follow the contractor to ensure that all systems related to provisional sum and PC rate have been incorporated in the electrical load, schematic diagram, and load schedules.

4. **Earthing system and lightning protection.**

In case of low-level water tables, earthing pits for transformers and RMU will be on the raft earthing system; this should be shown clearly in BOQ as the main contractor's scope of work.

5. **Mockup.**

Mockups to be prepared for installations as required by the consultant and the mockups to be checked and approved by the consultant before proceeding with the work were available.

6. **Basement design**

The consultant should follow the contractor to study the weight of all heavy equipment located on the ground floor above the basement, like the generator, MDB, pump, transformer, etc.

The contractor should ensure that the provided slab is designed to withstand the above-mentioned weights.

7. **Generator room.**

The consultant should follow the contractor regarding the generator room louvres that should be selected to fulfil the supplier's calculations for generator cooling, generator room is to be approved by the supplier regarding the generator requirements, which should be as shown below:

a) The room should be of enough size for the generator.

b) The generator should be inserted from one side of the room, which is the radiator side or from the alternator side,

c) The louvers should be behind the generator to allow air current to flow from louvers to the generator to the radiator then out of the room. Hence cooling the alternator, the engine, and the radiator.

d) Louver size is to be calculated based on the heat generation calculations.

e) Cable routing should be studied based on the cable entry to the cable box (trench to be used for button entry and cable trays for top entry).

f) A muffler should be installed to let smoke be thrown out through suitable opening in slab or wall..

g) Suitable location of the diesel tank and suitable arrangement for oil changing requirements depending on generator size.

h) The generator room drawing is to be signed by the supplier specialist to ensure all installation requirements are provided.

8. **Office building and retail area.**

The consultant should follow the contractor to ensure and check the following:

a. The location of the electrical DB is to be on a suitable wall and not in a partition.

b. The required cable between the DB and the source (SMDB) is shown with details in the design drawing, BOQ.

c. The DB should include a breaker, ELCB, etc., according to the authority requirement.

9. **Expansion Joint.**

If there is an expansion joint in the project's slabs, ensure using suitable fittings (expansion joint coupler) for expansion joints regarding the PVC conduits in or under the slab. **See 1-1**

10. **Safety requirements.**

The consultant should follow the contractor to ensure strict following safety rules and ensure using 110V equipment in the work (not 220V) if applicable.

11. **Size of doors.**

Door dimensions (height and width) for electrical rooms, generator rooms, mechanical rooms, Authorities rooms or any other service room should be studied by the contractor and discussed with the related authorities or suppliers to suit the equipment size inside and to allow future replacement for the biggest part inside the room. Doors are not to be installed by the contractor before installing all internal equipment.

1-1 PVC Expansion couplers to be used in buildings with expansion joints to allow expansion due to heat.

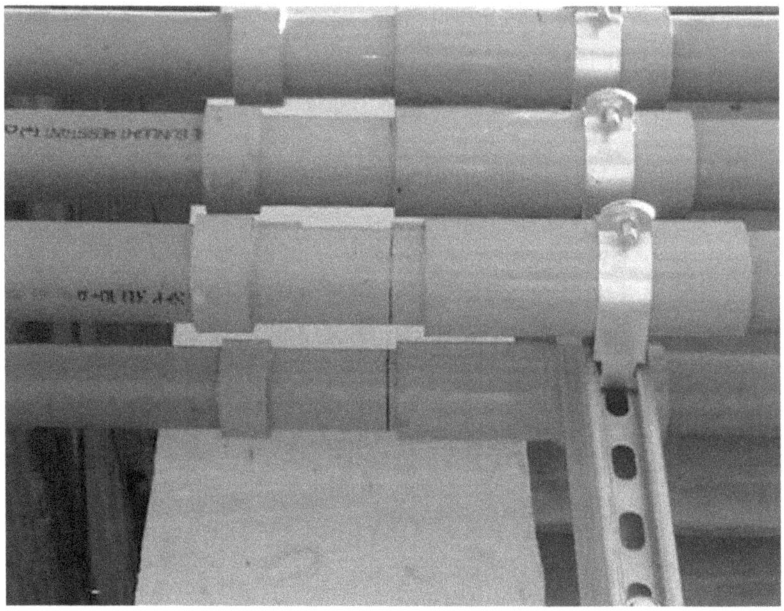

Note: some doors contain an Access Control System; hence, it is to be coordinated with the Access control supplier for his requirement regarding the provision for containment for electrical wiring and the type of Ironmongery that can be communicated with The Access control.

12. **IMS intelligent meter.**

The consultant should follow the contractor to check and ensure that the containment of the Intelligent Management system (IMS) KWHmeter and water meter with the required wirings and the process and components to transfer information to the control room are included in the design drawings, bill of quantities (BOQ) and specifications.

13. Garbage Chute.

The consultant should follow the contractor to check and ensure that the containment of the Garbage Chute (PVC or GI) is included in the design drawing, Bill of quantities (BOQ) and specifications.

14. Lighting control System.

The consultant should follow the contractor to check the followings are available in the lighting control system in the design drawings, Bill of Quantities, and specifications:

a. Override switches' location for lighting control.

b. The switches are to be outside the DB and near it in the electrical room.

15. GSM system

The consultant should follow the contractor to check and ensure that the cable tray and power supply for GSM system are included in the design drawing, Bill of quantities (BOQ) and specifications.

16. Double height (Void).

If there is a double height (Void) over the false ceiling, then the contractor should check if there is a suitable catwalk over the false ceiling to enable easy future maintenance for electromechanical work over the false ceiling and ensure that such requirement exists in the drawings, BOQ and the specifications.

17. Conduits in wet areas

The consultant should follow the contractor to check the following:

a. No PVC conduits under the floor in wet areas (bathrooms, toilet, kitchen, and any other wet area).

b. No DB installation is recessed on the wall of a bathroom or wet area.

c. No electrical room or electrical panel were installed under a wet area (bathroom, toilet, or kitchen).

d. No piping or wet service in the electrical room over electrical equipment. (Pipes, AC drains, etc.).

e. AC system or split units are to be away from the Switchgear, and a drip tray is to be provided under the AC system to collect the water to a suitable drain in case of water accumulation due to drainpipe blockage.

18. Lift, Lobby, Corridor, Balcony

Any surface conduit or open point over the false ceiling should be closed (both ends) by suitable sealing/ foam to prevent condensation inside the pipe due to the temperature difference over and under the false ceiling or around the light fitting in the balcony.

19. Temporary power supply cables

All temporary power supply cables are to be removed from the site after completing the project.

20. Location of tower crane

The contractor should ensure that the location of the installed tower crane in the project does not block the future work for the electricity or communication provider (or any other authority service) to provide their services at the end of the project.

Chapter 2

Starting the project

What is required for the project to start?

The consultant will direct the assigned Electrical contractor to start the study and follow up with the main Contractor or the design consultant for the design drawings that he will use to generate the workshop drawings. He should consider and check the following requirements in the life span of the project to ensure work quality and to control work progress complying with the project specification and scope of work:

- Contract Document:
- Organization Chart
- Communication tool:
- Periodical Meetings
- Breakdown of Bill of Quantities.
- Supervision Files:
- Drawing Packages:
- Quantity of shop drawings' sets to be approved by authorities:
- O&M Manuals:
- MEP Coordinator
- Submission to the Authorities
- Material Submission Reports Requirements.
- Drawing Submission Requirements.
- Flow Chart.

❖ Contract Document:

The following contract documents should be available on the site and should be studied carefully for any discrepancies, the following points to be considered: -

1. Points to be considered in Good for Construction Drawings and Tender drawings for Electrical and Mechanical Systems. The following points are to be studied and checked.

   a) Compare the "Good for Construction" and "Tender" drawings and list any discrepancies or differences. The contractor should discuss the changes or discrepancies found in the two sets of drawings. He should prepare himself for these changes, the required modifications to implement these changes, fees, and project duration expansion due to these changes. The above discrepancies or changes are to be reported to the resident engineer for legal rights, cost, and time implications.

   b) Make sure that "Good for Construction Drawings" includes all related authorities' approvals.

   c) Compare the quantities on drawings with the Bill of Quantities, taking into consideration adding any addendums issued to the contract if there are any.

   d) Check if all the technical points in the electrical specifications are complied with the design drawings.

   e) Coordinate with the mechanical engineer (contractor) to review electrical requirements for mechanical services on mechanical drawings. The contractor should ensure that all mechanical work requirements from the electrical side are included in the electrical drawings. The mechanical load should be the final approved load and based on the approved mechanical calculations.

   f) Check if the power required for the provisional sum and PC rate materials are considered in the drawings (electrical schematic diagrams, load schedules…etc.).

2. The Specifications are to be studied by the contractor to have a general idea about the project requirements.

3. The BOQ and the following addendums related to the contract (if available) are to be studied.

4. The contract agreement is to be studied for the scope of work.

5. The provisional sums and Prime Cost Rates are to be studied.

6. The soil investigation report indicating the water table in the site plot to be checked and to study its effect on the electrical installations.

7. The contractor should receive a copy of all authorities' approvals for design drawings related to fire alarm systems, emergency lighting systems, electrical (power and lighting systems), and communication (Telecom system and GSM).

❖ Organization Chart

The organization chart for the project's skeleton is decided depending on the project size. In small projects, the main contractor can handle all work, in medium or big projects, the main contractor will handle the civil and construction work. Hence, there will be a specialized MEP subcontractor for Electrical, mechanical and plumbing work, some projects will need a specialized contractor for the HVAC, landscaping, kitchen appliances, lifts, swimming pool, etc., depending on the project size.

On signing the Contract by the Main Contractor, the organization chart for the project skeleton will be submitted, including the MEP Sub-contractor who will handle the Electrical work. The organization chart for the Electrical contractor should consider the following:

1. Based on the project size and type, a specific organization chart for project teams should be agreed upon by the client.
2. The electrical engineer's attendance at the site should be decided whether full-time or part-time.
3. Confirmation that the electrical subcontractor is ready to submit all electrical material and get approval within a certain agreed duration (Example:1 or 3 months).
4. Confirmation that the electrical subcontractor will prepare and obtain all approvals for the workshop drawings within a certain agreed duration (Example: 6 months).
5. The Main Contractor shall confirm the MEP coordinator will be provided on board as a full-time job in the project (depending on the project size, it may require a coordination department).
6. CVs of Electrical subcontractor teams involved in electrical work, like Seniors, Electrical Engineers, site engineers, Supervisors, Foremen, etc. and their replacements in their annual leaves are required for review and approval. Replacement of Approved staff is not acceptable unless a similar experience is provided and they should pass the consultant interview.
7. Plan for replacement of contractor's engineers in case of annual leaves, resignation, or termination for any reason to be submitted to the consultant for approval.
8. The contractor's Valid Trade License is required to be submitted.
9. The approval of the proposed Electrical Site Engineer depends on his performance at the site and successful passage for the probation period.
10. The acceptance of the Proposed Electrical Sub-Contractor is subject to complete set-up with all necessary facilities, which shall include but not be limited to the following:
    a. Competent technical staff (managers, Engineers, Supervisors, skilled/non-skilled labourers, logistics etc.)
    b. Machinery, tools, instruments, etc., as required and deemed necessary for the proper execution of the electrical works.
    c. AutoCAD facilities include draftsmen, plotters, printers, photocopy machines, etc. or any other drafting tools like BIM.

d. Experience with similar projects.

11. The consultant will have the right to withdraw the conditional approval at any time if the performance of the electrical contractor and the quality of works do not comply with the expected quality, specifications, and instructions or if the work progress is not satisfactory.
12. The electrical subcontractor is responsible for obtaining the relevant approvals from the local authorities prior to carrying out the relevant electrical work and handing the same to the concerned authority after the completion of the work.
13. (If required) the testing and commission of all MEP work shall be carried out by an independent commissioning company as per the conditions of the contract.
14. All electrical works shall be carried out as per the approved materials, specifications, standard regulations, and electrical workshop drawings in accordance with contract documents and standard acceptable workmanship.
15. As per the contract agreement, at least One Electrical Supervisor shall be available on the site full-time for the period of construction.
16. The consultant should instruct the electrical contractor to start studying the project documents (Good for construction drawings, BOQ and the project specifications) and send all discrepancies, unclear details or contradictions to the consultant for review and updates.

❖ Communication tool:

Based on project specifications, official communications can be done through mail, fax or through specialized websites or platforms for construction projects like Aconex.

❖ Periodical Meetings

Periodical Technical meetings are to be started on the site for follow-up, coordination, discussions, etc., between the project's stakeholders, the consultant should attend the meeting shown below and any other meetings that require the consultant's attendance. That may affect project progress like:

1. Main contractor meeting with the consultant and the client (can be attended by the electrical sub-contractor and even some of the suppliers if required). This meeting is to discuss the progress of work; it can be done monthly or even every 15 days.
2. Senior Electrical Engineer meeting with Upper level (Main Contractor). This meeting is for discussion of work progress, planning for the electrical work, and coordination with civil workers and other contractors.
3. Senior Electrical Engineer meeting with suppliers of electrical systems. The meeting is mainly for material and O&M submittals to discuss the process of approval, workshop drawings, execution of works at the site,

and finally, testing and commissioning for installed systems. This meeting should be done with suppliers of most of the systems that require special expertise, as shown below:

1. **Low Current System**

    a) Gate barrier
    b) Electric door/sliding door.
    c) MATV system
    d) CCTV system
    e) Intercom, Security intercom and video call intercom
    f) Card access control system.
    g) Structure Cabling system
    h) Fire alarm system.
    i) Emergency lighting system.
    j) Sound system/Public Address system
    k) Garage door
    l) Central clock and bell system
    m) Automation System
    n) Nurse call system
    o) Foot counter system
    p) Parking guidance system
    q) Any other special or low current systems.

2. **Power System**

    a. HV system distribution network.
    b. Power transformers.
    c. Main, Sub-Main and Final Distribution Boards.
    d. Capacitor banks
    e. Generator and Automatic Transfer Switch.
    f. Motor Control Panels
    g. Busbar duct.

   h. Earthing and lightning protection system

   i. Lighting control system

j. Uninterruptable power supply system.

k. Generators.

l. Battery chargers.

3. Coordination with Authority (Technical Meeting)

   a. GSM system (Telephone system provider).

   b. Telephone system (Telephone service providers)

   c. Electricity Authority. (Electricity provider).

   d. Fire Alarm and Emergency lighting systems (Civil Defense).

   e. Radio communication (Security or Related Authority).

   f. Aircraft light (obstruction light) Civil Aviation Requirements.

   g. Excavation for cables (All related authorities in the area).

   h. Gas system. (Related Authority).

4. Coordination with Civil/Architecture Works (Technical Meeting)

These meetings are to discuss and distribute updates, additions, changes, latest approvals, design changes and approvals between all stakeholders and authorities. It discusses the following coordination activities between electrical, mechanical, civil, landscape and architectural work.

   a. Electrical work (switches and sockets) with furniture layout.

   b. Electrical work (switches and sockets) with tiles layout in kitchen and baths.

   c. Electrical work, Earthing and lightning pits with the landscape.

   e. Electrical work (DB installation recessed in walls; the wall should be not less than 20cm thick).

   e. Electrical work (panels in electrical rooms) with other systems' panels and equipment.

   f. Electrical work with mechanical services over the false ceiling.

   g. Electrical and mechanical work over bathrooms' ceilings.

   h. Electrical power supply for mechanical systems.

   i. Electrical services openings from the corridors to the rooms.

   j. Lightening protection with Civil columns, or façade (Architecture) and raft in the basement.

k. CCTV camera location with furniture layout and architectural work.

l. Telephone and MDF rooms with civil and architect.

m. False ceiling openings (service panels) with electrical services over the false ceiling.

n. Window cleaning system with civil and Architectural work.

o. Garbage chute system with civil work.

p. Lift (elevator)/ Escalators and travelators with civil work.

q. Health club and gym with civil and architectural work.

r. Swimming pool and Jacuzzi with civil and architectural work.

s. Sauna and steam bath with civil and architectural work.

t. Outside cables and Road/street lighting with landscape and infrastructure.

u. Duct road crossings with infrastructure.

v. Electrical component distribution with False ceiling tiles.

w. Glass rooms (bank entries) and related electrical and CCTV components.

x. Ground Trenches in Generator room, substations, and electrical rooms with civil work.

y. Cable/ services entries to the building with beams (civil work).

z. Cable risers and bus duct with civil work (slab opening).

aa. Generator room with civil and mechanical work (louvers, door, oil pit, water, fuel tank etc.).

bb. Roof work (MATV dish antenna) with architect (roof view).

cc. Underfloor trunking with civil work.

dd. Floor boxes with furniture layout.

ee. Access control system with Ironmongery and doors.

ff. Outside lighting, façade lighting on façade and ground (coordination with civil, architect and landscape).

gg. Communication protocol coordination with all involved systems in the loop.

hh. Machine Control panel and motor control panels.

ii. BMS input/ output signals for all systems according to project specifications

jj. Any other systems that are required to be coordinated with other work or outside clients.

❖ Breakdown of Bill of Quantities.

The consultant should follow the Electrical Site Engineer (contractor) to start preparing breakdowns for all items of the Bill of Quantities in detail, taking into consideration materials cost, labour cost (installation cost), testing and commissioning, overhead and profit for the purpose of processing/ approving the Contractor Payments.

Consultant and contractor should agree on the percentages of work and the percentages related to the equipment components' installation to enable the consultant to calculate the correct payment fees based on the agreed percentage criteria. Example for lighting point price: 30% for containment, 30% for wiring and 40% for fixing the light and switch.

❖ Supervision Files:

The consultant should follow the Electrical Site Engineer (contractor) to start preparing supervision files. The files should be organized as shown below: -

File 1. Design Checking: To list all design discrepancies.

File 2. Authority approvals: List the authority comments, corrections, meetings, and the status of the approvals.

a. Electricity provider: Anything related to the electricity provider requirement.

b. Telephone service provider. (Telephone). Anything related to communication provider requirements for landlines, GSM, cable TV, Etc.

c. Civil Defense. Anything related to civil defence requirements.

d. Civil Aviation. Anything related to civil aviation requirements.

e. Any other authority requirement (oil and gas field, army, security, environment dept., etc.).

File 3. Sub-contractor approvals. To list the status of the subcontractors' approvals for their systems.

File 4. Work progress meetings (Minutes of Meeting)

File 5. MEP Technical meetings (Minutes of Meeting)

File 6. Workshop drawings approval requests (covering letters and approval requests).

File 7. Material approval requests (covering letters and approval letters).

File 8. Project electrical load updates. All updates in load schedules and drawings.

File 9. Supervision checklist. All updates in supervision of equipment (monitoring and control).

File10. Inspection lists. To list the inspection requests, the location, the system, the date, the status, and no.

File 11. As-built drawings approval requests (covering letters and approval requests).

File 12. O&M manuals approval requests (covering letters and approval requests).

File 13. Feedback list. Any feedback from any stakeholder.

❖ Drawing Packages:

The consultant should follow the contractor to start working with the workshop drawings.

The workshop drawings are the detailed drawings that show the installation details to the level of the site foreman's understanding and requirements. It is created based on the design drawings and all following amendments. These drawings should be approved by the consultant and the authorities to be used in the installation work on the site. The drawings will include but not be limited to:

a. Containment drawings.

b. Lighting system

c. Power

d. Fire alarm system

e. Telephone system

f. Lightening protection system

g. BMS system

h. Emergency Lighting System

i. Low current systems

j. MEP Coordination drawings

k. Single-line diagrams and riser diagrams

l. Schematic diagrams

m. Installation details.

n. Services layouts

o. Builder's work

p. Coordination drawings.

❖ No of shop drawings' sets to be approved by authorities:

a. Drawing set for Consumer Department of Electrical Power Authority.

b. Drawing set for Distribution Department of Electrical Power Authority.

c. Drawing set for Telephone Service Provider Authority.

d. Drawing set for Civil Defense Authority.

e. Drawing set for Civil Aviation Authority. (In case of high tower projects).

f. Any other authority has jurisdiction in the location of the work.

❖ O&M Manuals:

Although the manuals are required in the last stage of the project, it is a good idea that the consultant informs the contractor to prepare the manuals and agree with him on the template of the manuals and the content he should provide so as to be approved as fast as possible.

In the final stage of the project span, all installed Systems manuals are to be provided and submitted for approval. The manuals should concentrate on the technical operation based on the site philosophy of operation and design, a table for the installed equipment with ratings and location of installation, the required preventive and corrective maintenance, spare parts, catalogues, special tools, troubleshooting, and warranty should be included in the manual. All manuals for the installed systems to be provided as an example (but not limited to):

a. Fire alarm system

b. Central Emergency lighting system

c. Stand-by generator system.

d. Central lighting control system

e. Sound/public address system

f. Structured/cabling system.

g. Closed circuit TV system.

h. Security system (Access control and security intercom system)

i. UPS system

j. MATV and/or IPTV systems (Internet Protocol TV)

k. Dimming system

l. Miscellaneous systems

m. Switchgear

n. Busbar

o. Interphone System

p. Gate Barrier System

q. Swimming pool

r. Lighting System

s. Facade lighting system

t. Lifts and escalators.

u. Any other systems.

❖ MEP Coordinator

The consultant should follow the main contractor in assigning an MEP coordinator for the project. The coordinator should follow and coordinate site activities between the main contractor, subcontractors, suppliers, and the authorities. He should ensure the spread of any updates or approval to all involved contractors, consultants, and affected suppliers to ensure integration between system installations. The MEP coordinator's duties and responsibilities should include the following:

At the Project Mobilization Stage:

The MEP coordinator should coordinate with project companies for MEP activities at the mobilization stage (installation of site offices) regarding the drawings for the offices, the approval of the drawings with the authorities, and the installation of the services for the offices on the site.

At the Project Execution Stage:

1) Review the MEP program (plan) and schedule of submittals to select the priorities of submittals according to site activities.

2) Follow up/ review all deliverables from MEP Subcontractors such as IRs, RFI, drawings, material submittals, closure of NCRs, and design changes. Etc.

3) Follow up approval of MEP submittals from the Consultant/ also to check all drawings and submittals before forwarding to a consultant.

4) Receive the results from the consultants for submittals, RFIs, design changes, and addendums and send them to all related contractors and stakeholders. Follow the stakeholders for their feedback that affects the drawings or the selected material.

5) Liaise with Consultant, Architect and MEP Subcontractor to finalise builders' work.

6) Follow up/assist in the execution of builders' work in accordance with approved drawings.

7) Arrange the Consultant's inspection for MEP work and follow up for the corrective action required to close the comments.

8) Evaluate/ verify MEP Subcontractor's variation claim.

9) Follow-up preparations/ submissions of redline markup drawings (for all concealed services), as-built drawings and operation and maintenance manuals by MEP Subcontractors. And following up on the

consultant's replies to the above submittals and spreading them to all stakeholders to ensure the required modifications in work.

10) Follow up with the subcontractor for tracking authorities for permanent utility connection.

11) Assist the Project Manager in monitoring the Subcontractor for removal and closure of snags in snag lists or punch lists.

12) The MEP coordinator is responsible for conducting a weekly coordination meeting with all subcontractors. A copy of the minutes is to be forwarded to the consultant.

13) The MEP coordinator shall monitor the site attendance of all MEP staff, ensuring the required staff for proper plan progress for site work by the contractor.

14) MEP coordinator shall be responsible for making necessary coordination of Electrical Systems work with Architecture, Civil, Structural, and False Ceiling Heights, finishing levels, landscaping, and finishes, which are essential for producing MEP shop and coordination drawings.

15) The MEP coordinator should study the contract documents. Any discrepancies found in any systems or in coordination with other systems, trends, or equipment should be pointed out to the Engineer at an early stage to allow corrective action.

❖ Submission to the Authorities

The Electrical consultant should follow the Contractor to prepare a flow chart for all Systems' activities related to the Authorities (Civil Defense, Electricity Provider, Telephone service provider, etc.) to submit the following requirements that are related to the mentioned Authorities and to prepare for the submission of the shop drawings for authorities' approval during construction. These activities are to be recorded in the related Supervision files or to be distributed to the concerned parties and stakeholders as shown below (but not limited to):

1. List of workshop drawings that are required to be approved by the Authority.
2. The method statement clarifies the steps required to obtain approval from the Authority for drawings, materials, NOCs (No objection certificates), etc.
3. The method statement clarifies the steps required to obtain both temporary (for site offices) and permanent power supply, required kWh, connection system for (24x7), and signal for communication system.
4. A list for coordination and workshop drawings which are related to the following Authority rooms:
    - Main and sub-electrical room
    - Main and sub-telephone room
    - GSM room

- Transformer room
- Ring main unit room
- HV switchgear room
- Pump rooms (mechanical rooms)
- ETS and BTU meter rooms.

5. Activity chart and program of work for all rooms related to the Authority shall be submitted clarifying steps for work execution (civil work, electrical work, mechanical work, finishing work, defining the contractor scope of work, etc.)
    a. Transformer and ring main unit room drawings.
    b. HT switchgear room drawings.
    c. MDF room and sub-telephone room/GSM room drawings.
    d. Main and sub-electrical room drawings.
    e. Firefighting pump room drawing.
6. All workshop drawings should be studied. Authority comments should be rectified for the authority's approval and then to be submitted for the consultant's approval.
7. Usually, the specification of the project requires the contractor to follow the latest updates in the applicable standards. Hence, a copy of all updated regulations/standards should be available and well-known to the contractor.
8. Flow chart for all switchgear materials (MDB, SMDB, DB and capacitor bank) showing installation location, type of fixing (surface or recess), actual load, design load, size of cables, type of cables, and drop voltage calculations, etc., to be provided by the contractor for for consultant's approval before submission for the authorities' approval.
9. List of all systems connected to the fire alarm system in addition to cause-and-effect Matrix as per Civil Defense Authority requirement to be submitted for manufacturer/ supplier specialist signature then to the consultant for approval.
10. All NOCs related to Authorities should be renewed at the right time by the main Contractor/ Consultant.
11. A key plan for the project should be provided, clarifying the location for all Authorities' rooms (transformer room, ring main unit room, main and sub-telephone rooms, firefighting pump room, all mechanical rooms, and all other service rooms) for authorities' approval.
12. Ensure that there aren't any drainage or water pipes passing through or above the authority rooms. No Authority room is installed below the wet area (bathroom, kitchen, or toilet).

13. Ensure that the shaft /riser opening between floors for electrical and telephone systems is with a stand around the shaft opening and threshold in front of all doors of the Authority room to avoid water flooding.

14. Ensure that the workshop drawing indicates the finished floor level (FFL) inside all authority rooms and outside as per Authority requirements.

15. Shop drawings are to be signed by the suppliers of the material. The signature gives an indication that the supplier specialist is accepting the installation details and that these details match the supplied system requirement.

❖ Material Submission Reports Requirements.

The electrical Consultant should instruct the Contractor to submit materials' submittals for the consultant's approval considering the following requirements:

1. Check and review the Good for construction/ workshop drawings, specifications, bills of quantities, table of finishing, etc., prior to proposing a material.

2. Material submittals shall include the following:

    a. Serial number and revision for the submittal.

    b. Introduction about the project (especially if the project is huge and there are several phases, the introduction will explain the project and its divisions or phases).

    c. The purpose of the submitted material.

    d. All components and equipment in the submitted system and locations to be used.

    e. Manufacturer details, supplier details (addresses and contact person) and country of origin.

    f. A valid supplier trade license with an expiration date and Ministry of Trade certification as a legal supplier of the material in the submittal. (To ensure availability of spare parts in future from the same supplier and technical support in case of failure.)

    g. Product data sheets highlighting the selected material with part number.

    h. The original Catalog highlights the used material with part numbers and the technical points required in the project specifications.

    i. Compliance statement to project specifications showing reference to the applicable catalogue page.

    j. Back-up calculations where required.

    k. Exact application and area/ location of use.

    l. Sample board.

m. Warranty for the proposed materials for the specified duration in the specification.

n. Detailed sketches.

o. Testing certificates.

p. Communication protocol (if it exists) ensures compatibility with other systems in the communication network.

q. A block diagram shows the system blocks and all connections and interfacings with other systems, components, and devices with the required power supply.

3. Material submittals should be in 4 sets; one set must have an original marked catalogue, and the other three sets can have photocopies of the selected/proposed materials. (Unless the submission will be through the network and the internet).
4. Each material (system) should be submitted in a separate material transmittal sheet.
5. All proposed materials should be from the specified manufacturer's list.
6. Materials from manufacturers other than the specified list are not acceptable unless a logical reason and justification are stated and agreed upon with client approval received.
7. The consultant shall reserve the right to save costs if an alternative material is proposed by the contractor and approved by the consultant.
8. All materials delivered to the site shall be inspected and approved by the Consultant prior to starting installation.
9. Materials should not be ordered unless final approval of the same is obtained from the consultant and authority (if required).
10. All electrical equipment, fittings, etc., should comply with the project's environmental conditions.
11. All materials, works, etc., should comply with codes, regulations, and requirements of local authorities for the systems such as:

    a. Power distribution system

    b. Fire alarm system and Emergency Lighting System

    c. Telephone system

    d. Aircraft washing lights/obstruction lights.

12. Materials should be ordered within 7 days from the date of approval.
13. The sample room should be available at the site to store and present samples for all approved materials with proper references and signatures by the project manager and consultant for records and future

reference. It is a good practice to encourage installing the samples to their respective components, like installing light fittings to the sample of false ceiling and switches and sockets to a sample of the wall with finished paint to match the colours, the signature of the client is mandatory to approve the material that is deemed to be installed in the front of the house.

14. Flow chart for the submitted materials highlighting the specified part numbers, proposed materials, name of manufacturer, supplier, country of origin… etc. shall be submitted.

❖ Drawing Submission Requirements.

- **Workshop Drawings**

The electrical consultant should follow the Contractor to start the preparations of the drawing procurement schedule, electrical shop drawings, detailed drawings, and flow chart, taking into consideration the following comments.

1. The drawing procurement schedule should contain the drawings related to all systems in the project, the drawings to be submitted based on the most critical drawings that are required in the early stage, and then the next, based on the project progress, the most required drawing are as shown below:

    a. Sleeves and side entry for services

    b. Containment in RCC (reinforced concrete columns).

    c. Power containment and recessed socket boxes in the floor, walls, and columns.

    d. Low current systems containment in the floor, walls, and columns.

    e. Earthing and lighting systems (sleeves for earth pit cables only).

    f. Lightening System in columns (if lightening strips are inside slab and concrete columns).

    g. Builders work for the required openings in the slabs for the risers in electrical rooms and the required openings in the walls for horizontal services.

    h. Lighting containment in the GF ceiling.

    i. Low current containment in the GF ceiling.

    j. Power containment in GF ceiling for Mezzanine floor power.

    k. Low current containment in GF ceiling for Mezzanine low power.

    l. Conduits in walls for GF for lighting, power, and low current separately.

The above sequence is to be repeated for the first floor, second floor, etc.

    m. Riser diagrams for all systems (power and low currents).

    n. Cable trays and trunking containments for each system on each floor

o. GI conduits installations on each floor (if applicable) for the back of the house.

p. Technical room details.

q. Sections and sketches.

r. Special area details (Entrance, lobbies, roof, meeting rooms, etc.).

2. General Points to be Considered in Workshop Drawing:

MEP Subcontractor is responsible for preparing Workshop Drawings containing all information deemed necessary for proper execution at the Site:

a. Installation details drawn in proper scale, Layers, Representations, Levels, Dimensions, Spacing, Sections, Type of Slab (Post Tension, Hurdy, Concrete) with Expansion Joint (if exist), Type of Wall (Block or Concrete), Type of Finishing…etc., should be mentioned in the Workshop Drawings.

b. Ensure that the Dimensions of the Lift Pit match the approved Lift material (consult the lift supplier engineer for his requirements for lift services; his signature is mandatory on the drawing).

c. Workshop Drawings should clarify the names of all rooms as per Architectural Drawings. The names of the rooms should be the same in all trades (mechanical, electrical, architectural, etc.). The same should be done for staircase numbering, hall numbering or names, flat numbering, and the same identification of numbering systems to be common between mechanical and electrical drawings ..etc., so the same name or number for the location or equipment should appear in all drawings, hence for any discussion, all parties considering the same location.

d. The location of the Switch and Socket should be coordinated with the direction of the door, door frame width and furniture layout. The set of switches, thermostat, fan regulator and any other accessory should be in the same order in all installations, the same to be done for the sockets, telephone, data, TV, etc.

e. Separate Workshop Drawing should be submitted for the Telephone System, Fire Alarm System, Emergency Lighting System, Power, Lighting…etc.

f. Any workshop drawing related to a specialized supplier (fire alarm, central batt. System, BMS, Lifts, escalators, travelators, lighting control, Access control, CCTV, etc., are to be sent to the supplier specialist engineer for his signature. The specialized engineer's signature means the drawing is correct and matches the manufacturer's requirements for installation.

g. The Workshop Drawing should indicate the Finished Floor Level (FFL), Drop Beams, Dimensions of all Areas, and Type of Wall.

h. All Workshop Drawings should have space at the lower right corner for Consultant Comments and Stamp for Signature/Approval.

i. All symbols used in the drawing shall be described in the Legends Column.

j. 4 numbers of Workshop Drawings should be submitted for approval, one for the Consultant (Client) Office, One for Consultant's Site, one for the Main – Contractor and One for the Sub – Contractor) unless the submission is through the Internet.

k. A tabular column showing the mounting height for all wiring accessories, DB, isolator…etc. In different locations, it shall be part of the legend column in the workshop drawings. If the legend is not enough, the contractor should submit the mounting height schedule for the consultant's approval, and the schedule should be updated for every new requirement on the site.

l. A tabular column showing numbering for all flats, shops, electrical rooms, telephone rooms, service rooms, pump rooms, offices, stores, fire alarm system devices, and emergency lighting systems. etc., shall be included in relevant workshop drawings. The mentioned flat numbers or room names should be the same in all trade drawings.

m. One set of the latest Approved Workshop Drawings (with Approval Stamp) shall be available on the Site for execution of works. Any changes on the Site due to Site Condition/ Consultant Comments shall be marked on the Drawings for As-Built Drawings' preparation.

- **Detailed Drawings**

The electrical consultant should follow the Contractor to start the preparation of the detailed drawings after approval of the workshop drawing, taking into consideration the following:

a. Coordination Drawings:

1. The electrical consultant should instruct and follow the Contractor to start the preparation of the coordination drawings list after approval of the workshop drawings, taking into consideration the following requirements (but not limited):

   a. RMU (ring main unit) and High-tension rooms with details (if they exist).

   b. Main and sub-main rooms (Low tension)

   c. Main telephone and sub-telephone rooms.

   d. Transformer room

   e. Main entrance (Especially for lobbies in hotels or tower entrances).

   f. Main and sub-garbage rooms

   g. All mechanical rooms (pump room, AHU Room, BTU meter room… etc.)

h. Operation room, control room and BMS room

i. Lift machine room

j. Kitchen /Tiles layout (for coordination with electrical installations).

k. Reflected ceiling layout and activities above false sealing.

l. Shaft layout

m. Furniture layout/flat

n. Generator room

o. All service areas around the building, including earth pits.

p. Parking Areas

q. GSM rooms

r. Lightening protection coordination with the facade. And lightening. Protection copper conductors with steel structure drawings.

2. Coordination shall clarify the following (but not limited):

   a. Coordination between electrical installations (wiring accessories, lighting fixtures, ...etc.)

   b. Coordination between electrical and Mechanical installations (A/C system, grills, diffusers, water heater, thermostat, access openings...etc.)

   c. Coordinate between electrical and civil work for the required openings in slab for risers, and openings in walls and beams for trays and bus ducts and ground lighting distribution.

   d. Coordination between electrical and Low current systems (fire alarm devices, emergency lighting, access control, CCTV, and intercom...etc.).

   e. Adequate sections and section details to clarify the heights and distances between systems.

   f. Coordination between electrical work and false ceiling tile inside the kitchen and areas with rectangular false ceilings or gypsum board ceilings.

   g. Coordination of electrical work with architectural and landscape work for lighting installation, socket in tiles, power supply for admin desk and kitchen cabinets, and façade and landscape lighting.

   h. Any other required coordination based on site conditions.

Note: Full coordination can be done using BIM technology by CAD drawings using Revit.

b. Key Plan Drawing:

The key plan shall clarify the location of the following service rooms:

a. Hi, Voltage Switchgear, Transformer and RMU rooms (if they exist).

b. Main Electrical and Submain Electrical Rooms.

c. Main Telephone, Submain Telephone and GSM Rooms.

d. All Mechanical Rooms (pump, Exhaust Fans, BTUmeter and AHU Rooms).

e. Main Garbage Chute and Sub – Garbage Chute Rooms.

f. Generator Room.

g. Lift / Elevator Machine Room and escalator pits.

h. The main entrance and other sub–entances.

i. Health Club and Gym Area.

j. Gate Barrier / Electric Door / Sliding Door.

k. Electrical substations.

c. Builder's Workshop Drawing:

The electrical consultant should instruct and follow the contractor to prepare the Builder's Workshop Drawings after approval of Workshop Drawings and Coordination of Workshop Drawings indicating the following details (but not limited):

a. Wall openings with dimensions for all penetrating services (Cable Trays, Trunking, Bus Duct… etc.).

b. Slab openings with dimensions for all penetrating services (Cable Trays, Trunking, Bus Duct… etc.).

c. Final Location of all equipment with dimensions.

d. Location of entry points for all Services to the Building (Inside and Outside the Building), ex power cables entry, telephone connection entry, outside lighting or fountain/ swimming pool cables, etc.).

e. Water Level inside the Plot.

f. Finishing Floor Level [FFL] and height of False Ceiling [reflected FC].

g. The slope for all Service Rooms, Electrical Rooms, Telephone Rooms… etc., on all floors to ensure no water accumulation towards these rooms.

h. Drop Beam with dimension (if any).

i. Two Sleeves with suitable sizes inside all lentils shall be considered for any future requirements.

d. Block Diagram:

The electrical consultant should instruct and follow the contractor to prepare a Block diagram for approval with the submitted workshop drawings for the Electrical systems in the project (Some are shown below) to clarify the way of connection between subunits and interfaces with other systems:

1. Gate barriers.
2. Electrical door/sliding door with access control system.
3. Fire alarm system (interlock with all other systems)
4. Mechanical equipment: -
    a. Water supply pumps (CP) control panel.
    b. Sump pumps CP
    c. Firefighting pumps CP.
    d. Transfer pumps CP.
5. Intelligent metering system (IMS)
6. Car Parking management system
7. 24x7 (Direct Alarm System) connections to Civil defence.
8. Lighting control system.
9. CCTV system
10. Security intercom system
11. Access control system
12. BMS
13. Public address system
14. Door Interphone system
15. A/C Thermostat connection.

e. Elevation and sketches:

The electrical consultant should instruct and follow the contractor to prepare the required details for elevations, clarifying the installation details and dimensions for all equipment coordinated with other services for the systems shown below (but not limited):

1. Gate Barrier.
2. Electrical Door / Sliding Door with Access Control System.
3. BTU / Water Meter Room.
4. Generator Room.
5. BMS Room, Control Room and Security Room.
6. Main and Submain Garbage Chute Room.
7. All Mechanical Rooms [Pump Room and A/C Room].
8. MDF Room, GSM Room, Telephone … etc.
9. Main and Submain Electrical Room.
10. Lift Machine Room.
11. Parking Management System.
12. Main Entrance and Lift Lobbies.
13. Public Toilets.
14. Kitchen Tile Layouts.
15. Façade lighting.
16. Side Wall lighting in the staircase (to be coordinated with staircase step levels).

f. Riser and Single line Diagram:

The electrical consultant should follow the contractor to prepare Riser and Single Line Diagrams after approval of the Workshop Drawing for the activities shown below (but not limited):

1. Parking Management System.
2. Emergency Lighting System.
3. Telephone System.
4. Power System.
5. Fire Alarm System.
6. MATV System.
7. CCTV System.
8. Intercom System / Security Intercom System.
9. Card Access Control System.

10. Public Address System / Sound System.
11. Lighting Control System.
12. Lightning Protection System.
13 Uninterruptable power supply system (UPS).
14. Computer network (MSN).
15. Any other system depends on project specifications.

g. Isometric Diagrams:

The electrical consultant should instruct and follow the contractor to prepare the isometric diagram for special systems like bus ducts, garbage chutes (if there are bends in the chute), and vacuum transmission systems in hospitals to draw the correct lengths, bends, angles, and components matching the manufacturer dimensions in the technical catalogues. Such a diagram will require the following information to be collected for correct drawings:

1. The exact dimensions of the approved system from the catalogue with its accessories (bus duct, Garbage chute, vacuum transmission).
2. The exact approved Finish floor level for all floors is taken from the approved civil/ architectural drawings.
3. The exact approved floor height for all floors is taken from the approved civil/ architectural drawings.
4. The exact location for the openings for the riser (openings to be in the same place to ensure a straight riser to the last floor unless there is a change in the raiser location then the system should be steered in a suitable location (under the false ceiling for the bust duct) to the new opening location. (The duct route in front of the house should contain a false ceiling so the duct can be installed over the false ceiling).
5. The exact dimensions of the system room. The bus duct will require the dimension of the flange or adaptor to the electrical panel (or transformer) and the plug-in unit with the related cable size and the required bend that will supply each floor.

h. Wiring Diagram:

The electrical consultant should follow the contractor to prepare and provide the Wiring Diagrams after approval of Workshop Drawings for the activities shown below (but not limited) to clarify the way of operation, installation, connection, etc.:

1. Two-Way Switch / Intermediate Switch / Dimmer Switch.
2. Staircase Control Switch.
3. Emergency Lighting (Self-contained, sustained, central battery).
4. Thermostat [FCU, CU, PS, Relay …etc.] A/C System.
5. Lighting Control System.
6. Intercom and Security Intercom System.
7. Parking Management System.
8. Access control system.
9. Uninterruptable power supply system.
10. SMATV system (panels/DB & MDU's).
11. Control panels.
12. Required power supply for fire alarm interfaces.
13. All Switchgear (HV, LV, PF corrections).

i. Calculations

The electrical consultant should instruct and follow the contractor to prepare all required technical and electrical calculations for the project as mentioned in the project specifications or as required by the consultant to ensure correct sizing, the following calculations are to be provided but not limited to:

1. Voltage drop calculations (4% drop is accepted from the MDB to the final DP).
2. Short circuit calculations.
3. Coordination study.
4. Power factor correction system.
5. Current and voltage harmonic distortion.
6. Battery charger sizing.
7. Battery sizing for UPS.
8. Lux level calculations.
9. Loading calculations for cable trays and bus duct supports.
10. Any other calculations required based on project specifications.

❖ Flow Charts

The consultant should instruct and follow the contractor's Electrical engineer to start preparing the flow chart after completing the approval of workshop drawing and details drawing for the following activities (but not limited to) to clarify in detail the breakdown of the system, program of work, material procurement schedule, relation with Authority, drawings etc...., and to simplify the way of follow-up for the project.

1. The contractor mechanical Engineer should be responsible for providing a list of actual mechanical equipment loads like (FCU, CCU, AHU, FAHU, exhaust fan, extract fan, booster pump, sump pump, transfer pump, firefighting set pumps ... etc.). The sizing of mechanical equipment should depend on the contractor's specialized engineer/ designer to verify the design and submit the final correct ratings to the contractor's electrical engineer for supply sizing.
2. The electrical work program should be submitted. Priority of work should be dedicated to the Authorities' spaces/rooms, pump rooms, telephone rooms, electrical rooms, generator rooms, RMU, and GSM. Room... etc.
3. MEP Subcontractors shall submit their method statement for work execution, securing the quality of the installations.
4. Material procurement schedules and drawings flow sheets shall be submitted showing all necessary data, including proposed submittal dates; these sheets are to be updated biweekly.
5. MEP Subcontractor shall prepare a method statement showing all necessary steps for obtaining the final Authority's approval (Electricity, Telephone, Civil Defense, Civil Aviation, etc.).
6. The contractor shall submit a list of equipment/ systems to be connected to the fire alarm system indicating the description, used protocols, interface type, application and type of indications required.
7. The contractor shall submit a list of equipment/ systems to be monitored or controlled by BMS (input-output matrix) at an early stage for proper coordination, clarifying whether the BMS is for monitoring or control and monitoring. The contractor should check the protocols for all systems connected to the network are equal and can be communicated easily. BMS suppliers should be consulted to verify the list.
8. The contractor shall prepare a breakdown BOQ for billing purposes, i.e., percentage of material, installation, testing, etc., for payment interim calculations.
9. The contractor shall submit the mounting height schedule of all wiring accessories and devices according to the specifications, regulations, and client requirements.
10. Schedule for false ceiling clearance (all false ceiling areas, rooms, type of false ceiling ... etc.).
11. List of switchgear schedules to be submitted (MDBs, SMBs, DBs, emergency switches, gear panels, transformers, RMUs, FPs, LPs, control panels, isolators ... etc.).

12. Finishing for all walls externally and internally shall be submitted.
13. Numbering for all rooms, flat service rooms, etc., shall be submitted.
14. An overall activity chart for the whole project shall be submitted.
15. A schedule for lighting fixtures shall be submitted, including locations, type of installation, rating, type of bulb, and thermal temperature.
16. The contractor must submit for approval samples of all types of engraved labels for all electrical equipment devices and junction boxes, etc.
17. A schedule of CCTV system details shall be provided showing CCTV location, camera type, circuit number feeding the CCTV cameras, and Type of suspension (bracket/ceiling or on column).
18. Schedule for lifts, escalators and travelators details showing the type of equipment, size, location to be installed, isolator and ELCB ratings.
19. A list of all Main and Sub-electrical rooms with reference names/ numbers (if applicable) shall be provided.
20. Schedule for all control panels in the project showing (control panel name, location, bracket on wall/ skid on pump, canopy/ normal, IP degree of protection, connected load, feeding location … etc.).
21. Schedule for all activities related to low current systems like (all required provisions, power supply, interface with other systems, panel location … etc.).

Chapter 3

Drawings

In this chapter, the project's drawings will be clarified:

- Shop and coordination drawings.
- Builder's work drawings.
- As-built drawings.

Shop and coordination drawings.

Preparations for shop drawings

Shop drawings are generated from design drawings or good-for-construction drawings. The design drawings are created by the designer. Design drawings are general drawings that show the systems in the site in general without any installation details or dimensions for a specific brand or manufacturer. Hence, when the contractor procures the electrical systems, these systems may need certain requirements for installation; they may have certain dimensions that differ from other systems and may have some additional components or some connections or interlocks to other systems. All these requirements are not shown in the design drawings as the design was made in general without any material selection or coordination with other systems.

When the contractor selects, gets the approval, and procures the material, he should start preparations for shop drawings that are being drawn based on the selected systems. It shows the exact dimensions for the selected systems and installation details to the level of the details required by the technician in the site. It should show the practical entry and exit for cables and wires, the coordination, and interlocks with other systems. These shop drawings will show the exact installation method that should be implemented in the site. The shop drawings are required to be approved by the consultant and the related authorities. The consultant should instruct the contractor for the followings:

1. Preparing the shop drawings submission plan and arranging meeting invitations to the main contractor with all subcontractors and system suppliers to discuss and study the shop drawings submission plan and preparation for workshop drawings.
2. 5 sets of workshop drawings to be submitted by the contractor for approval as follows (unless project communication is done by Internet):
    a. One set for the consultant
    b. One set for the main contractor
    c. One set for electrical subcontractor
    d. One set for the client
    e. One set is to be on-site for the preparation of as-built drawings to mark all changes that occurred on the site due to site conditions and the progress of work (especially above the false ceiling).
3. All workshop drawings related to low current and other systems should be sent to all suppliers' specialists for stamping and signature. Suppliers' specialists' signatures are mandatory as confirmations that all system requirements for correct installation and functioning are provided.
4. Workshop drawings related to the Authorities should be approved by their respective Authority with consultant approval.

Schedule for Workshop Drawing Preparations

Based on the project size and specifications, the following shop drawings are required to be submitted for approval. The type and details of electrical and low current sytems in the site will give indication about the amount and the required shop drawings required to be submitted for consultant's approval. The schedule of submitted shop drawings should include (but not limited to) the followings:

1. Lighting system
2. Power system
3. Fire alarm system
4. Telephone system (data)
5. Lightening protection system
6. Earthing system.
7. Emergency lighting system
8. Low current system
    a. MATV system
    b. Intercom system
    c. Gate barrier system
    d. CCTV system
    e. Parking management system
    f. Power management system
9. Riser and schematic diagrams for power distribution system and load schedules.
10. Riser diagram for all low current systems (CCTV/Access/Fire Alarm/EM lighting/MATV, etc....).
11. Any other system based on project specifications.
12. Sketches and details.
13. Submitted drawings should include the following details.
    a. Proper scale.
    b. Finishing floor level inside and outside the building (pavement around the building).
    c. Sections and details (scale 1:20).
    d. Block wall and concrete Thickness.

e. Internal and external finishing types are to be marked on the drawing.

f. Flow chart clarifying civil items related to electrical works like (Internal and external finishing type, false ceiling type, etc..........)

g. Type of slab (post tension/normal/precast/hurdy) and expansion joint (if exist).

h. Clear and enough dimensions.

i. Names or identifications of all areas and rooms based on architectural drawing, taking into consideration electro-mechanical rooms' numbering, store, etc....

j. All drop beams with dimensions.

k. Inverted level for all activity related to outside services.

l. All comments from the Authority during the design stage are to be considered in the workshop drawing.

m. Space for the legend, including symbols used and general notes.

n. Bulkheads in false ceilings inside rooms should be shown to allow correct lighting distribution.

o. The water level inside the plot should be marked on the drawing.

p. The mounting height for all wiring accessories, DB, isolator, etc., is to be shown either on the drawing or submitted for approval as a schedule of mounting height.

q. Space at the lower right corner for consultant comments and stamp for signature/approval.

Note: Shop drawings are to be submitted for each floor unless some floors are typical then a suitable keying system is required for flat and location numbering.

Schedule of special or technical room drawings:

Based on the project size and specifications, the following technical and special room coordination drawings are required to be submitted for approval. These drawings should include all systems, builder's work, architectural and civil details for authorities' approval. The submitted room details and shop drawings will be based on consultant's and authority requirements that may include (but are not limited to) the following.:

1. Main and sub electrical room, transformer room, ring main unit room.
2. Main and sub telephone room/MDF room/GSM room.
3. All mechanical rooms for air conditioning systems.
4. All mechanical rooms for plumbing.
5. All service areas (Car parking areas, lift lobby, corridors, etc....).

6. The external area around the building.
7. Toilet's ceiling coordination drawing.
8. Roof, lift machine room, garbage chute room, etc.…
9. Generator room.
10. Kitchen layout with elevation tile, including all applicable services.
11. Health club/gym room and swimming pool.
12. BMS room/control room.
13. Shaft layout.
14. All elevation lighting was coordinated with the architect's work.
15. Coordination drawing for all activity related to reflected false ceiling layout.
    a. For areas with false ceilings, show the final location with dimensions for all accessories and devices, lighting fixtures, FAS devices, sprinklers, grill diffusers (supply and return air grills), CCTV cameras, access opening and all other related items, etc.…
    b. Show all activity above the false ceiling in detail to coordinate the space between the false ceiling and all other services above, duct routing, drainage or water pipe routing, fan coil units, lighting point, and all other low current points, etc.…
16. The furniture layout for the flat is for coordination with other services.
17. Coordination drawing should include plan, section, elevation, etc.…., clearing all activities related to electro-mechanical work and other trades, taking into consideration Authority and supplier requirements.

Points to be considered before starting preparations for shop drawings/ coordination drawings and details drawings, etc.…

1. Study and coordinate the location of the power supply, control panel, supporting system and interlock with other systems for all A/C equipment (condensing unit, FCU, FAHU, AHU, chiller system, grills, diffuser, thermostat, ex-fan system, jet fan, pressurization fan and all electrical operating valves etc.).
2. Study and coordinate the location of the power supply, supporting system and control panel for plumbing equipment (central heater system, swimming pool, Jacuzzi, sump pump, water feature, sauna and steam system, water booster pump, water transfer pump, etc.) to be finalized.
3. Study and coordinate the location of the power supply, supporting system control panel and interfacing with other systems (firefighting system, CO2 system, foam system (FM200), sprinkler system, etc.).
4. Study and coordinate the location of power supply/ control/ supporting system and interlock with civil

and mechanical systems in drawings for window cleaning (cradle), garbage chute, smoke management system, air curtain system, irrigation system, gas system, etc.

5. Study and coordinate the location, power supply, and interlocks for low current systems (Fire alarm, CCTV, emergency lighting, etc.). Also, check the communication protocols to ensure smooth communications between interconnected systems.
6. Study and coordinate the location of all power points, lighting points, lighting fixtures, switches, sockets, Isolators, and other accessories.
7. Study and coordinate the location of switch and socket in coordination with direction of door opening, width of door frame, and furniture layout.
8. For Lentil – Provide two sleeve or as required in each lentil above the door for switch, thermostat (taking into consideration direction of door opening … etc.)
9. All points in reinforced concrete areas (walls, columns, slab) should be marked in advance on shop drawings as these points should be prepared in early stage in the steel reinforcement before closing the shutter for concrete pouring.
10. Dimensions of the lift pit should match the approved lift materials, manufacturer requirements and workshop drawings.
11. Before starting the workshop drawing, ensure to refer to the final updated good for construction for the following drawings:
    a) Structure drawing.
    b) Architecture drawing

Note: The support system for the plumbing and AC equipment is designed and installed by mechanical team, coordination with mechanical team for the location of the supporting channels so as not to block opening the equipment doors for maintenance purpose, and not to reduce the walkway width and prevent future equipment replacement.

Builders Work Drawings

Builders workshop drawings:

Builder's work is the construction drawings that show the openings, sleeves in the slabs and the walls for electro-mechanical services. These drawings should be prepared to show the locations where the electro-mechanical services (ducts, cable trays, openings for risers, etc.) will go through the walls and slabs, coordinated with the related shop drawings.

The consultant should instruct the contractor to prepare builders' workshop drawings after the consultant and authority approval of the workshop and coordination drawings, clarifying and ensuring the following:

1. Builder's workshop drawings like (transformer room, ring main unit room, HT switchgear room, MDF room, main and sub-telephone room etc.) should be approved from related authority and consultant for the required wall and slabs openings with dimensions for used cable trays, trunking, busbars, ducts etc.
2. Builder's workshop drawing for electro-mechanical rooms (main electrical room, pump room, AHU room, FAHU room, chiller area, water feature, landscaping … etc.) should be approved by the consultant for the required wall and slab openings for cable trays, trunking, busbar, duct, pipes etc. See 3-1

3-1 Slab opening for electrical services in the Electrical room.

3. For areas like (the authority room, roof floor, basement parking area, mechanical room, and electrical room) which have drop beams, suitable openings or sleeves should be provided for all services for services' penetration like (cables, cable tray, sprinkler pipe, drainpipe, bus bar etc.) taking into consideration provisions for all mechanical services. See 3-2

3-2 sleeves are in the beam for piping, and MEP services are provided in the concrete stage.

4. For (authority room, roof floor, basement parking, mechanical room, electrical room, etc.) which have opening between floors or walls, suitable sleeves or openings should be provided for all services with (5-10%) extra space to pass the applicable services like (cable, cable tray, bus bar, air condition duct etc.).
5. Authorities-related rooms like (telephone room, MDF room, ring main unit, electrical room, transformer room, etc.) should be far away from any wet areas around or above; otherwise, attic slab and double wall should be provided as per standard and Authority approval.
6. For high water levels or the possibility of a water leak, the contractor should use a puddle flange, especially for sleeves and ducts passing through the wall of a water tank or basement sheer wall; the puddle flange is to be shown in the drawings. All sleeves and ducts should be sloped to prevent water from going inside the building.
7. The finishing floor level should be marked on the drawing (inside and outside the building). Also, the height of the false ceiling is to be marked on the drawing.
8. All underground ducts inside the building should be marked on building drawings.

As-Built Drawing

As built drawings are the final bunch of project drawings. It will be drawn by updating the shop drawings to show the electrical systems in the site as built practically by implementing any change in the site installation due to any reason.

For services over false ceilings, contractors usually mark a set of shop drawings (Called markup drawings) to show the practical installation for the servicers over false ceilings and to be signed by the consultant for the record when being inspected and approved. This will enable the production of as-built drawings in the future without opening the false ceiling to re-check the routing of the service.

The as built drawing should show the electrical services in the site as built exactly with all details and information required for future maintenance. It should be produced with several copies as required in the project specifications. It should cover all services in the site with all installation details, risers, coordination's, interlocks, input output matrices, schedules, legends etc.

Chapter 4

Coordination with Other Services and Mechanical Work

The most important work that requires special and accurate coordination is mechanical work. Mechanical systems and related equipment require an electrical power supply for operation. The quality of power installation work will affect the quality of the mechanical systems installation and their smooth operation through the life span of the project. Hence, the coordination between electrical and other systems, especially mechanical systems, is so critical to ensure correct installation is being implemented for all other systems, including mechanical systems and then correct and reliable operation.

The consultant should instruct the main contractor to start a series of weekly coordination meetings, including electrical and mechanical engineers (civil and architects can join if some systems are related to their work), electrical and mechanical suppliers, and all related stakeholders to discuss the electrical work required for smooth and reliable operation of all systems requiring electrical power supply. The meeting should cover installation and coordination work between systems, including the followings and to be recorded in the minutes of meeting and distributed for record:

1. Estimation and discussion of all systems requiring power supply, including mechanical systems, philosophy, and mode of operation, interfacing, and interlocking with other systems and the required control.
2. Provision of equipment schedule with all details of mechanical systems and components.
3. Agreement for the required submission plan for systems approval and approval requirements.
4. Mechanical contractor confirmation for the final sizing and ratings for mechanical equipment based on the latest sizing calculations. The final equipment is to be submitted to the electrical engineer for power supply consideration with mechanical consultant approval.
5. Deciding the priorities for the submitted drawings and material submittals.
6. Any unexpected events that required change requests or modifications.
7. Details of coordination work and activities between systems components and the requirements.
8. The communication protocol between systems (if applicable).
9. Progress follow-up for electro-mechanical work.

10. The replies to the previous meeting comments.
11. Any other issues?

In Addition, the consultant should instruct and follow the contractor for the following:

1. Submission of coordination drawings for all mechanical and special rooms with internal mechanical services showing the required sections, elevations, plans…etc. These drawings to be submitted showing the mechanical services and the related electrical services (with dimensions) coordinated with other mechanical, electrical, low current and civil systems for installation requirements for the ducts, cables, trays, pipes, beams, lights, sensors, dampers, pump and chiller bases and interfaces etc. The drawings are to be approved by the mechanical and electrical consultant for the following areas as shown below:
    a. Mechanical rooms (Pump room/ Ex-fan room/ Air handling room/ Chiller room/ Cooling tower area/ BTU room).
    b. All service areas (lift, lobby, entrance, etc.)
    c. Basement parking, etc.
    d. Roof
    e. Garbage room
    f. LPG Gas room
    g. Control room and BMS room
    h. Swimming pool area/ Jacuzzi/ health club/ Gym.
    i. Sauna steam bath
    j. MDF room
    k. Irrigation plant room
    l. Water feature
    m. Air Curtains
    n. Any other service room or any room contains systems that require an electrical supply or panel.
2. Flow chart for all electro-mechanical equipment mentioning the location of equipment, Electrical data (Supply rating, actual mechanical load, rating of breakers, isolator ratings, cable size and type), Interface with another system like building management system (BMS), Fire alarm system…etc. (The contractor should ensure a common communication protocol for all systems communicating through the network) as shown below.
    a. Chiller system and chiller management system

b. Fresh air handling unit

c. Fresh air fan

d. Chilled water pump

e. Package system

f. Split unit

g. Mechanical room

h. Pump room

i. Firefighting pumps

j. Water heater / Boiler

k. Water booster pump

l. Sump pumps

m. Water feature pump

n. Air handling unit

o. Control panels

p. Smoke exhaust fans

q. Jet fans.

r. Pressurization fans

s. Swimming pool and fountains

t. Water supply pumps

u. Water transfer pumps

v. CO2 detection system.

w. FM 200 System.

x. Smoke management system.

y. Any other mechanical or architectural system.

3. Activity chart (status of work) for all service rooms (pump room, AHU room, garbage room, BMS room…etc.) mentioning and indicating in detail the plan for the work execution for the next week or month (look ahead schedule) for (Civil Work, Electrical Work, Interface with other systems, finishing work, testing, and commissioning with monthly updates.

4. Block diagram for all activities related to each system to be submitted clarifying the connection between all activities, power supply and interlocks with other system (BMS, Fire alarm system, control panel, pump, float switches, dampers, interfaces… etc.).
5. Wiring diagrams for all systems and interlocks to be submitted, including power and control wiring, as shown below:
    a. Thermostat connection
    b. Water heater/ Boiler connection.
    c. Condenser and Fan coil units with pressure switch and thermostat
    d. BTU meter
    e. Sensors for window AC.
    f. Firefighting pump (all connections between pump, power supply, Fire alarm system., Zone control valves, flow switch…etc.)
    g. KWHmeter and water meter with connections required for intelligent meter system.
    h. Interface with fire alarm system.
    i. Water supply pumps, Water booster pumps, water tanks, float switches…etc.
    j. BMS with all other system interfaces and power supplies required.
    k. Kitchen hood (Switch, fan…etc.)
    l. Chilled water chemical treatment (power switch, control …etc.)
    m. Water feature and related pump.
    n. Jet fan
    o. Air curtain system
6. The consultant to follow the contractor to ensure the followings are available in the coordination drawings:
    a. Dimensions of all mechanical, electrical, control equipment and panels (MCCs, MCPs, pump sets … etc.) should be studied to fit within the dimensions of the mechanical room and can be inserted through the door and can be removed in future for replacement.
    b. The final location of control panels should be as per the approved workshop drawing and Authority approval. If the pump is installed on the underground floor, the Control panel should be on the ground floor, and a disconnecting switch should be near the pump to ensure power disconnection for safety at maintenance time.

c. Up-stand should be provided for all mechanical rooms' doors to prevent water entry to other areas in case of flooding, slope for finish floor level towards drain point should be considered.

d. Power, control cables and wires sizing between (MDB/SMDB) and control panels (Motor Control Centers and motor control panels) to the pumps or other services should be calculated based on actual mechanical load, distance from supply and installation method.

e. VFD (variable speed drives) installation should be done according to supplier specialist recommendations. Special attention to be given for the installation location, required cables and cable lengths. VFDs should have filters to reduce harmonics.

f. The final location of all equipment, devices and accessories related to mechanical systems should be according to approved mechanical drawings, authority approval and manufacturer recommendations ex (pumps, control panel, thermostat, fan coil unit., C.U., water heater, window A/C, routing of duct, water pipe, A/C pipe … etc.).

g. Check the availability of the A/C system inside all mechanical rooms.

h. For the District cooling system: The contractor should coordinate with the service provider for the required power and scope of work.

i. Power supply for the mechanical and other systems should be fed from separate ELCB+CB breakers (should not be linked with other systems) and sized for the actual mechanical load.

j. Confirmation from the supplier of all pumps/ motors for the following:

    i. Frequency 50 Hz

    ii. Temperature 50 C

    iii. Insulation of winding as per specification.

    iv. If motors are controlled by SS or VFD, then motors should be selected to match soft starting and VFD control.

k. If the cooling system is fed by the district cooling plant, then the drawings should be approved by the authority. Work should follow the authority requirements in addition to the above requirements.

7. Special attention should be paid to the support systems for mechanical equipment, heavy support, or steel channels fixed to the ground should be designed carefully for the location of installation so as not to block the maintenance work or reduce the walkway width and prevent equipment replacement.

Coordination with Mechanical Systems.

Below are some of the systems to be coordinated with electrical work (power supply, control, sequence of operation, communication protocol):

1. **Air Conditioning Systems:**
    a. Air-cooled chillers
    b. Water-cooled chiller (with cooling tower).
    c. Package unit systems.
    d. CU + FCU (split units)
    e. District cooling
    f. Window type
    g. CCU (close control units).
    h. Air handling units.
    i. Fresh air handling units
    j. Exhaust fans
2. **Fire Fighting System:**
    a. Sprinkler systems
    b. Foam systems
    c. FM 200 systems
    d. CO2 systems
    e. Zone control valves
        (1) Butterfly valves
        (2) Water flow switches.
        (3) Drain valves.
    f. Motorized dampers.
    g. Pre-action system
3. **Water Supply System:**
    a. Water supply pumps
    b. Water transfer pumps
    c. Sump pit pumps
    d. Irrigation plant rooms

e. Swimming pool systems

Coordination of Mechanical Systems with Fire Alarm System.

Below are some of the systems to be coordinated and interlocked with fire alarm systems based on civil defence regulations and requirements:

1. FM 200
2. CO2 system
3. Air conditioning systems
    a. Fresh air handling units
    b. Air handling units.
    c. Motorized dampers on supply/return A/C duct
4. Aspirating system – VESDA (smoke detection system)
5. Smoke and Pressurization Fans.
6. Lift Pressurization Fans
7. Basement Exhaust Fans
8. Firefighting system:
    a. Firefighting pumps
    b. Jockey and diesel pumps
    c. Firefighting Diesel generator.
9. Sprinkler systems:
    a. Sprinkler butterfly valve
    b. Sprinkler flow switch
    c. Zone control valve
    d. Fire hose cabinet
    e. Alarm test station
    f. Check valve.
10. Smoke Management System:
    a. Smoke extract fans
    b. Pressurization fans

   c. Motion fire smoke damper.

11. Foam System Control Panel.
12. Air Curtain System
13. BMS
14. Garbage System

Water tank

The consultant should follow and check the contractor to fulfil the following activities before and during the installation of the water tank:

1. No conduit work is allowed on the floor of the water tank; no panel is allowed to be installed on the wall of the water tank.
2. Front view elevation of the water tank should be submitted by the Contractor with all details clarifying the following:
   a. Waterproof (W/P) 13A S.S.O (switch socket outlet) at a high level outside the tank near the water tank opening for general service and future maintenance (above water level).
   b. Waterproof (W/P) Junction box 20cm x 20cm at high level above water level near the window for connection of float switch between control panel and water tank (The box to be installed from outside water tank wall with back sleeve from rear of box to tank for sensor installation).
3. The contractor should submit a block diagram including connections between all water tanks and float switches to a control panel that controls the operation of water pumps in the tower or site fed from underground, ground, and roof water tanks. The drawings should show all components related to the tanks and all connections to all interfaced systems.

   Note: Special care is to be taken for the services' installation in concrete water tanks, as any missed service or mistake in the installation can't be repaired because no chipping is permitted in the water tank, which will lead to future cracks and water leakage.

Motor Control Center (MCC).

Material submittal requirement

The consultant should follow the contractor to provide the material submittal report for the Motor Control Center, the contractor should provide the following for the consultant's approval:

1. The mode of operation of all Motor control centres (MCCs) should be submitted by the Mechanical Engineer for the consultant's approval before submitting any workshop and schematic drawings for the motor control centre.

2. Compliance statement to be submitted highlighting and marking the complied and non-complied items, with justifications for the alternative items if not complied.
3. Type of motor starting (VFD, DOL, star-delta starting, soft starter ... etc.) should comply to the project specifications, authority regulations and supplier or manufacturer recommendations.
4. Workshop schematic drawings to be included in the submittal for consultant's approval complying with the approved mode of operation, including all support documents (catalogue, calculations) clarifying the following:
    a. Cable, bus bar, wire sizing.
    b. Rating of breakers, fuses, contactors, relays ... etc.
    c. Motor starting type (DOL, star-delta, VFD, soft starter ... etc.).
    d. Terminal block diagram for all connections showing incoming and outgoing wires and cables between the control panel and all interfaced systems like BMS, FAS, Float switch, AC system, water supply ... etc.
    e. Voltmeter, ammeter, selector switch and all other accessories.
    f. Low voltage transformer for the control circuit.
    g. Mechanical interlocks for star delta starters, if applicable.
    h. Phase failure and under voltage relays.
    i. ELCB/ELR for each pump (should be 30mA tripping sensitivity).
5. Flow chart for all MCCs related to mechanical equipment clarifying the following: location, actual load, isolator rating, breaker ratings in all related panels, cable entry and exit, cable sizing, SMDB reference, Control panel reference, interface with other systems... etc.
6. The type of enclosure, like (an IP rating canopy for all external control panels, depends on the installation location of MCC
7. Make sure that the disconnecting switch or emergency push button is provided for each pump or near the door of the pump room where applicable.
8. Enough ventilation is available in the MCC room or MCC panel, and the number of ventilation fans is enough.
9. A detailed maintenance plan or program for the first-year maintenance to be submitted by the contractor for approval, this plan to be updated yearly to the handing over stage. The plan should be submitted to the client to continue the maintenance activities accordingly.
10. A list of all components related to MCC to be provided showing the manufacturer, local supplier,

ratings, part no for MCC components, and definition of symbols Etc. This to be generalized to all installed systems to facilitate future spares procurement.

11. Electrical wires used in control panel wiring should be off for a 105°C rating.
12. IP rating for the enclosure to be checked with respect to the project design and to the final location of installation.
13. A detailed drawing/ sketch for the float switch connection for water supply pumps is to be provided in the submittal.
14. VFD protection function to be clarified.
15. All circuit breakers shall be compensated for 50°C in the electrical room and 55°C in the mechanical rooms.
16. Protection components of the panels should be from a single manufacturer for compatibility.
17. Motor protection should be done by using motor protection type circuit breakers or contactors, over current and under voltage relays should be provided.
18. The rating of all components should be written on a single diagram and should be consistent with the approved pump and chiller.
19. The size of the enclosure should be adequate to avoid wire congestion, difficulty in maintenance and temperature rise. The type of enclosure should be as approved drawing or submittal with 2mm steel thickness.
20. Vertical bus bars for outgoings are to be in separate compartments.
21. Based on the contract scope of work, the contractor should submit a maintenance schedule and spare parts list for the duration specified in the project specifications.
22. Ensure coordination with a mechanical engineer for the procured pumps that may contain overheating protection sensors in the winding. A suitable control circuit should be provided in the MCC for such protection. Connection details are to be submitted and clarified.

Installation of MCC

The following points should be checked before and when installing the MCC:

1. MCC should be assembled as per the requirements of the concerned Authorities (Electricity provider and civil defence). The MCC should be inspected and tested at the factory in the presence of the Consultant and Contractor's Engineer before delivery to the site.
2. The contractor must submit an installation method statement based on the supplier's recommendations; the MCC testing and commissioning method statement, including standard readings and the success

criteria, should be submitted before T&C.

3. Make sure the size of the MCC matches its dimensions in the workshop drawings.
4. All components in the MCC or any other panels should comply with the approved material submittal.
5. Make sure that all electrical connections for bus bar, cable and wires are tightened robustly according to the correct torque, bolts should be marked after torque.
6. Make sure that the identification engraved labels for all devices are provided (breakers, fuses, relays, contactors, starters, timers ... etc.), ferrules also should be provided for all power and control wires.
7. Make sure that the rating and setting for all breakers, contactors, timers, VFD starter, relays, fuses etc. are calibrated as per the approved coordination study, approved workshop drawings of MCC and as per the actual mechanical load.
8. Make sure that the operation of all breakers and other accessories is mechanically functioning correctly (on /off/ trip).
9. Make sure that the earthing system is completely connected.
10. If the MCC is located underground or in locations difficult to access, or open area, the MCC panel should be equipped with a warning alarm hooter on the panel or a disconnecting switch or emergency push button on the ground floor near the door or, where applicable and be connected to BMS.
11. The MCCs should contain LED indicator lamps for (ON/OFF and trip) status for each pump.
12. Make sure the provision of a disconnecting switch or emergency push button for each motor near the pump room is connected to the MCC.
13. Check the operation and functioning of the anti-condensation heater.
14. Make sure that all live parts are suitably shrouded.
15. Keys are to be fixed near the control panel using the suitable key box.
16. MCC power supply should be fed by a separate feeder breaker if the MCC require an emergency supply in addition to the ordinary supply. Both supplies (normal and emergency) of the MCC should not be shared with any other control panel.
17. For MCCs fed by emergency power, ensure time delay for the supply of the emergency power when changing over from normal to emergency power.
18. Testing and commissioning method statement to be submitted including the successful test readings and a blank format to be filled with the test results.
19. ELCB sizing is to follow the Authority requirement and common practice, and the location of ELCB is to be either in SMDB/ DB or in MCC.

20. All bus bars shall be fully insulated, and RYB colouring should be identified.
21. Final approval is subject to the successful testing and commissioning of the MCC (motor control centre).
22. Make sure no water pipes are passing above the MCC.
23. Make sure the location of MCC is in an accessible area, and as per the approved coordination drawings, the location of MCC and control panels should not be on the underground floor. Refer to the Authority regulation.
24. Check that the routing of cables, wires, and control wires between the MCC panel and equipment is coordinated with all other services and designed for the entry/ exit of the cables in the MCC.
25. Ensure all doors leading to MCC or mechanical rooms are wide enough to pass all electro-mechanical equipment like (MDB/ SMDB/ MCC/ PUMP... etc.) to the room through the door.
26. Make sure that all mechanical rooms are conditioned by the A/C system.

Control Panels

Materials Submittal Approval for Control Panels

The consultant should follow the main contractor to arrange periodic meetings between the electrical engineer and the supplier of control panels to discuss and clarify the philosophy and mode of operation, specifications, and installation requirements. The contractor should be notified to provide the followings:

1. The Mechanical Engineer should submit all control panels' Modes of Operation for consultant approval before submitting the workshop drawings for control panels.
2. The type of starter (VFD, DOL, star-delta, soft starter ... etc.) should be according to the specification and authority regulations. The starter should be coordinated with the supplier of the control panel.
3. A compliance statement to the project specification is to be submitted highlighting the complied and deviated items, with justification for deviations.
4. Full catalogues for the used material with part numbers in the catalogues highlighted.
5. Flow chart for all control panels related to mechanical equipment clarifying the following: location, actual load, distance between the panel and the load, isolator rating, breaker rating, feeder breaker rating in SMDB, cable size, cable entry and exit, SMDB reference, CP reference, interface with other systems and BMS etc.
6. A detailed maintenance plan or program for the first-year maintenance is to be submitted by the contractor for approval; this plan is to be updated yearly to the handing-over stage. The plan should be submitted to the client in the handover stage to continue the maintenance activities accordingly.
7. The size of the enclosure for the control panel should be large enough to avoid wiring congestion,

future difficulty in maintenance, and internal temperature. Enclosure steel thickness should comply with the specifications, and usually, it is 2mm thick.

8. Ingress protection IP rating and canopy for the control panel should be suitable for the installation location (Internal and external).

9. Make sure that the ventilation of control panels is adequate and the number of ventilation fans are enough.

10. A list of all components related to MCC or control panel is to be provided, showing the name of the manufacturer, local supplier, the definition of symbols... etc.

11. Control panel wiring shall be suitable for 105°C.

12. All live parts must be shrouded.

13. All protection components inside the control panel should be from the same manufacturer.

14. The control panel interfaces with other systems like fire alarms and BMS systems (if applicable).

15. All circuit breakers shall be compensated for 50°C in the electrical room and 55°C in the mechanical rooms.

16. All bus bars shall be fully insulated and RYB colour identified.

17. Motor protection type breakers are to be used for motor protection, with overcurrent and undervoltage relays.

18. Careful coordination with mechanical load is required for the following:

    a) Sometimes, the design of some AC systems (in some package units) requires more than one feeder for power supply (fans, compressors, and heaters); hence, it may require including the control of the AC system in the control panel.

    b) For panels supplying power for sump pumps, the pumps should be procured by mechanical engineer with long cable enough to reach the panel connection unit without any joint.

    c) For motors, in general, the cable entry to the pump and the gland should be studied and selected to match the power supply cable that may take a range of sizes based on the distance between the pump and the panel.

    d) Some motors contain heat sensors in the windings that required to be connected to the control panel for motor winding protection hence, the control panel should be equipped with such facility.

19. Control panels power supply should be fed by separate feeder breaker. Some panels may require emergency supply in addition to the ordinary supply. Both supplies (normal and emergency) of the control panel should not be shared with any other control panel.

20. Valid supplier trade license with expiry date, and ministry of trade certification as a legal supplier of the material in the submittal.

Workshop drawings for electrical services related to mechanical work:

The consultant should instruct the contractor to start the workshop drawings and schematic diagrams for the electrical services for mechanical work, showing the mode of operation and including all support documents, wiring diagrams, and calculations, clarifying the followings:

1. Cables and bus duct Sizing and wires.
2. Ratings of breakers, fuses, contactors, relays ... etc.
3. Starting method of pumps (DOL, star-delta, VFD, soft starter ... etc.).
4. Terminal block diagram for all connections showing entry and outgoing wires and cables between the Control panel and other systems like BMS, Fire alarm system, Float switch ... etc.
5. Voltmeter, ammeter, selector switches and all other accessories.
6. Low voltage transformer for the control circuit.
7. Mechanical interlock if the type of starter is star-delta.
8. Control and protection relays like phase failure and under voltage relay.
9. The earth leakage circuit breaker (ELCB) and earth leakage relay (ELR) for each pump should have a (30mA) sensitivity.
10. Detailed drawing for the water supply pump and the connections to the float switch.
11. Single line diagram should show the Ratings of all components and should be consistent with the other AC system components like the chiller and cooling tower.
12. The control panels drawn in the workshop and coordination drawings should match the practical size in the submittal for proper and correct coordination with other systems.
13. Interfacing with other systems like BMS.
14. Coordinate with civil and architectural work for entrance door dimensions for (pump and electrical) rooms; it should be enough to allow entry of all electro-mechanical equipment like (CP, PUMP, switchgear ... etc.) to their respective rooms.
15. For control panels located on an underground floor, in areas with difficulty to access, or in open areas, then the control panels should be provided with a warning alarm hooter on the panel. Disconnecting switch or emergency push button on the ground floor near the door or where applicable; these details should be shown in the drawings. Control panel to be connected to BMS.

Notes for Chillers

1. The LV power supply for the chiller should be sized according to the authority requirements, the sizing usually should be as shown below based on practical experience for the transformer's reliable operation:
    a. 1000KVA transformer shouldn't be loaded by more than a 650KW compressor.
    b. A 1500KVA transformer shouldn't be loaded by a compressor that is more than 800KW.
2. Load more than 100KW should be through multistage compressors with operation delay between compressors' starting of 1-3 minutes.
3. The starting method should be using soft starters.
4. Always consult the authority for the latest regulations.

Installations for electrical services related to mechanical work.

The consultant should follow the contractor to check and ensure the followings in the installation of the electrical services related to the AC systems and components.

1. The contractor should submit and get the consultant's approval for the installation method statement based on the supplier specialist's directions and recommendations.
2. Control panel should be assembled, inspected/ tested according to the Authority requirements in the factory in the presence of Consultant and Contractor's Engineer before delivery to the site.
3. Containment, cables, and wires between pumps and control panels should be installed based on the approved workshop drawings and coordinated with all other services.
4. Activity chart and program of work for control panel shall include steps for executing the following work: Civil work, Electrical work, Finishing work, Testing/Commissioning. The chart should define weekly or monthly look ahead activities with Monthly Updates.
5. Proper protection for all control panels, wires, junction boxes ... etc., should be provided during installation.
6. For panels fed by emergency power, ensure time delay for the supply of the emergency power when changing over from normal to emergency power.
7. The final location of all control panels should be as per the approved workshop drawing.
8. Ensure provision of identification and engraved labels for all devices (breakers, fuses, relays, contactors, starters, timers ... etc.). Ferrules are to be provided for all power and control wires.
9. No water pipes are allowed to pass over the electrical and control panels.
10. Ensure all electrical connections for bus bar, cables and wires are tightened robustly to the correct

torque value using torque spanner and marked by a suitable marker.

11. Ensure Air conditioning is available in the design of the mechanical and control rooms.
12. For control panels located in underground floor, in area with difficulty to access, or in open area, then the control panels should be provided with warning alarm hooter on the panel. Disconnecting switch or emergency push button in ground floor near the door or where applicable, control panel to be connected to BMS.
13. Keys are to be fixed near the control panel in a suitable key box.
14. Installation of the earthing system shall be as per supplier or manufacturer recommendations and the project specifications.
15. The contractor's Site Engineer should monitor the installation activity daily and continuously till completion.
16. All control wires and power cables shall be with extra length for easy connection to the equipment terminal.
17. The ELCB size and sensitivity rating should be as per Authority requirement, and the location of ELCB should be either in SMDB, DB, or CP based on the design.
18. Suppliers should propose the working staff details and names of the personnel that will be in charge of the project. Periodical or half-monthly site status reports shall be submitted by the supplier to the consultant and client for follow-up.
19. The breakdown for all items related to the installed systems shall be proposed by the Contractor for billing purposes; these breakdown percentages should be agreed upon by all concerned parties.

Thermostat

1. The contractor should submit the thermostat wiring within the submitted shop drawings showing the following:
    1) In and out of power supply for fan coil unit.
    2) Control wires between thermostat and fan coil unit.
    3) Location of the components (relay, thermostat, fan coil, solenoid ...etc.).
2. Flow chart for all FCU's connection types with all details and components, depending on the actual mechanical load.
3. The used thermostat is to be approved by a supplier specialist or manufacturer.
4. Make sure that the location of thermostat as per the final approved location in mechanical drawing and coordinated with other services and furniture layout.

Fan coil unit

1. The location of the power supply and junction boxes for the fan coil and the thermostat should be in an easily accessed area; and in the correct side of power connection to the fan coil.
2. The direction of the strainer should be positioned away from any distribution board/ lighting fixture and any power outlets above the false ceiling.
3. The location of the fan coil unit should be as per the approved coordination drawing (mechanical and electrical). The drawing should clearly show the direction of supply and return ducts, water pipes, and power supply outlets for the fan coil unit.
4. The rating of the fuse (in fused switch outlet) for FCU should be as per actual mechanical load.
5. A Flow chart for all fan coil units in the project should show and clarify the following:
    (1) Fan coil details
    (2) The installation location above a false ceiling or surface type.
    (3) Ordinary or decorative type.
    (4) Type of power supply.
    (5) Actual mechanical load.
    (6) FCU rely on, if required, a relay connection with the thermostat.
    (7) Location of power supply connection and water pipes entry/exit.
    (8) Classification of fan coils based on above specifications to facilitate selection of the correct fan coil for each location (Type 1,2,3,4 etc.).
6. Make sure the duct between the fan coil unit and grills has been coordinated with mechanical and electrical work to avoid covering electrical points or other services for future maintenance.

Building Management System (BMS)

Usually, there will be a specialized subcontractor for the BMS; based on the system type (monitoring or monitoring and control) and project specifications, the specialized contractor will create the input-output signals table for all systems connected to BMS based on the design and project specifications.

It is common in projects that some of the signals are missed in the specification. The BMS subcontractor usually advises to include, or due to some design changes, some additional signals should be included; hence, the final input output list that covers all required information from the BMS for the specified systems, whether for monitoring, or monitoring and control should be approved by the client and the consultant, this list may include variations due to added signals to the systems that need to be approved by the client.

Before starting the installation work, the consultant should instruct the contractor to consider the following points:

1) All systems connected to BMS must have a separate circuit/separate ELCB with an engraved label clearly showing the reference circuit and DB.
2) All systems related to BMS requiring low-voltage power supply should be marked on the power drawings.
3) Flow chart for all control panels, direct digital control (DDC) or PLC related to BMS, and all other components related to the systems clarifying the following:
    a. Power supply (circuit reference, DB, location, fuse rating with indicator lamp, etc.).
    b. Interlocks and interfaces with other systems like (Fire alarms, lifts, escalators, pump control panels, garbage chutes, AHUs, gate barriers, ACCESS control, water tanks, CCTV, etc.).
    c. Telephone point details and location (if applicable).
4) The scope of work for electrical should be studied, discussed, and coordinated between the BMS contractor, Electro-mechanical Engineer, and other suppliers for any interlocked or interfaced system to discuss installation requirements, especially the communication protocols.
5) A coordination drawing for the BMS room showing all activities inside the room should be submitted for approval.
6) Block diagram/ riser diagram should be submitted for approval showing conduit sizing/junction boxes/ type of wires and cables /power supply (fused with indicator lamp) / telephone point (type and wiring), any required relays … etc.
7) The sequence of operation should be approved by the mechanical Engineer for all mechanical systems.
8) Ensure that all identification with engraved labels and ferrules is provided for Electrical work.
9) Make sure the panels (DDC) related to BMS have the required power supply (fused with indicator lamp) and telephone outlet for data transfer. (As per supplier requirements).
10) All the above activities are to be checked and followed up during the fixing/ installation of the system.
11) Ensure that the installation of the earthing system is as per the supplier's recommendation.

Water Heater

The consultant should follow the contractor for the requirements shown below for water heater installation:

1. The power outlet for the water heater w/h should be in an accessible area.

2. Water heater installation to be coordinate with other services especially the water pipes to allow easy replacement for the heater element in future.
3. The location of the water valve related to the water heater is away from the electrical power supply or any electrical service.
4. The final location of the water heater should be as per the final approved coordination drawing (mechanical/ electrical).
5. Ensure that the installation of the earthing system is as per the project's applicable standard and supplier recommendation.

Swimming Pool, Jacuzzi, Sauna, and steam bath.

The consultant should follow the contractor to ensure the following requirements:

1. The Main Contractor must arrange periodical meetings between the Electrical, Mechanical Engineer and system supplier for installation discussions and provision of the installation requirements as follows:
    a) Flow chart for all equipment related to the system including the following data (actual load for mechanical equipment/ type of starter/ power supply data (MCCB breaker rating, size of cable)/ interlock and interfaces with other system (BMS, FAS, etc....)/ location/ control panel/ earthing system, disconnecting switch, isolator ... etc.) shall be submitted.
    b) Coordination drawing for pump room, swimming pool, Jacuzzi, sauna, and steam bath, including all activities like (light points, lighting switches, emergency lights, power points, control panels, cable routing, pumps, low voltage transformers, emergency stop push button, isolator, cable tray, etc....) to be submitted for approval.
    c) Block diagram for all activities related to the swimming pool, sauna steam bath and Jacuzzi for mentioning all connections between systems and interfacings like (light points and fixtures, LV transformer and drivers, pump Control panel, fire alarm, BMS, emergency push button, size of cables, earthing system, power points, float switch ... etc.) to be submitted to the consultant for approval.
    d) The scope of work for electrical services should be studied and discussed between suppliers of swimming pool/ Jacuzzi, sauna, steam baths and Electro-mechanical Engineer.
    e) All control panels for the system should be submitted for consultant approval after being approved by the mechanical Engineer for the mode of operation.
    f) The contractor should submit and get the consultant's approval for the installation method statement based on the supplier specialist directions and recommendations. Then a separate method statement

to be submitted for testing and commissioning for all electrical work.

g) Breakdown for all items related to the system (electrical work) for billing system to be agreed by all parties.

h) Activity chart and look ahead program of electrical activities related to the mechanical systems installation work to be submitted to the consultant defining the scope of work for each party with monthly updates (civil work, electrical work, mechanical work, finishing work, testing, and commissioning work).

2. The consultant should follow the contractor to confirm the following steps in the installation of the above-mentioned systems.

    a. No containment is allowed to be installed on the slab of the swimming pool. (can't be reached for maintenance).

    b. Identification with engraved labels and ferrules to be provided for all control panels and other electrical works.

    c. The location of the isolator and double pool switch (DP) should be coordinated with mechanical and architectural drawings.

    d. Cable sizing for underwater lighting should be studied taking into consideration distance between fittings, LV transformer, voltage drop, location of drivers. Etc.

    e. Additional length of one meter should be considered for underwater light cable to allow lifting the fitting to enough height for maintenance.

    f. Swimming pool lighting box installation should be coordinated with the system supplier and ceramic tiles work.

    g. Swimming pool, sauna, steam bath and Jacuzzi systems should be connected to their corresponding power distribution board or panel through separate MCB and ELCB with (10mA) ratings.

    h. Ensure that the installation of the earthing system is as per the project's applicable standard and supplier recommendation.

    i. The supplier will assign the authorized engineer for certification of the system.

    j. The supplier should submit the installation program with the look ahead schedule. The Specialized supplier Engineer should be continuously following the site activities to ensure quality.

3. Final approval shall be granted based on the successful commissioning of the system.

BTU meter Room

The consultant should follow the contractor to ensure the following requirements during the installation work of the BTU meter room:

1. All BTU meters' locations should be installed as per the service provider or the Authority of district cooling requirements and regulations (in the cabinet, above the false ceiling, at the entrance of the flat). Installation should be coordinated with other services for the followings:
    a. Discussion of Power for B.T.U. Meters are fed by built-in batteries, or they may need a Low-voltage power supply.
    b. Discussion of data transfer, whether it required a certain containment, or it will be through a wireless antenna, or by BMS network.
    c. Communication protocols are to be studied and coordinated with other interfaced or interconnected systems.
2. Detailed Coordination drawing for the BTU meter room should be submitted to the consultant for approval, including layout, sections and elevations for all walls, including the location of all activities as follows:

    BTU meter, lighting point, lighting fixture, switches, power supply for the systems, Fire alarm devices, conduit, trunking opening for A/C (grill), emergency light, telephone point, control wires to the (BMS), chilled water pipes, data wires for the systems.
3. Block diagram/ riser diagram for all activities related to the BTU meter system shall include BTU meter, power supply, BMS network, antenna, telephone system, junction box, containment and wiring required for the installation of the system between sub-BTU meter, cabinet, chamber, antenna, and main BTU meter if any.
4. The power supply for the system shall be from an emergency supply if a standby generator is available and from a separate circuit and ELCB.
5. Complete coordination between electrical, mechanical, and architectural work and the installed systems shall be done, clarifying the scope of each party's work.
6. BTU meter arrangements should be according to the service provider's regulations and requirements.

Intelligent Water Meter

The consultant should follow the contractor to ensure the following requirements during the installation work of the water meter room:

1. Coordination drawing in details for water meter room should be submitted to the consultant for approval including layout, sections, elevations for all walls mentioning the location of all activities Ex: (Water meter, lighting point, lighting fixture, switch, power supply for the system, Fire alarm system

devices, conduit, trunk, opening for A/C (grill), emergency light, telephone point if any).

2. Block diagram/ riser diagram should be submitted to the consultant for approval including all activities related to intelligent water system as follows (water meter, power supply, telephone point if any, junction box, containment's, wires' sizes, connection between sub-room and main room and chamber etc.).
3. The power supply for the system should be fed from emergency panel if stand by generator is available in the site and to be from separate circuit, section and ELCB.
4. Complete coordination between Electrical, Mechanical, and architectural work and the installed systems shall be done, clarifying the scope of works of each party.
5. Provision for intelligent metering system (IMS) should be confirmed as per the regulations or the requirements of the concerned Authority.
6. The water meter arrangement in the room should be as per the authority's approval.
7. Study the communication protocol with other systems like BMS.

Pump Rooms and Air handling rooms.

The consultant should follow the contractor to ensure the following requirements during the installation work of conduit, outlets, pumps, control panels, cable trays, cables, DBs, switches, etc. in the mechanical room:

1. The contractor must start periodical coordination meetings with all sub-contractors and suppliers working in the mechanical room to discuss and prepare a breakdown activity chart for all systems, civil works, A/C works, electrical works, etc.
2. Main Contractor to submit coordination workshop drawing to the consultant for approval showing all details with dimensions (plan, elevations and sections) location, routing of cables/wires, pumps, MCC, control panels, A/C duct, water supply conduits, drainage pipes, cable trays, trunkings, power supply boards, disconnecting switches, emergency push buttons when applicable (front of pump or near the main door for pump room), emergency light, lighting fixtures, light points, waterproof (WP)switches and sockets, firefighting equipment, fire alarm devices, BMS panels ... etc. (coordination drawing should be stamped by all sub-contractors and suppliers before submission for consultant's approval. The stamp of each contractor means he checked the drawings, and everything is shown matching his requirements, and there is no need for changes).
3. Distribution of lighting fixtures inside the mechanical room should be in coordination with the location of pumps, control panels ... etc. Also, the fittings should be waterproof and IP 65, bracket, or ceiling type.
4. All exposed or surface containment for cables should be on GI cable trays or GI conduits routed in straight lines (vertical/ horizontal).

5. The support system for the AC ducts should be shown in the drawings, especially the one fixed on the ground. The support base should be selected so as not to affect maintenance of any AC equipment by blocking door opening or reducing the walkway width that prevents equipment replacement.

Testing and commissioning for electrical services related to mechanical work.

1. The consultant should instruct the Contractor to complete all pending or delayed work related to mechanical systems. Testing and commissioning Method Statements to be submitted by the contractor for the consultant's approval including standard readings and the successful test criteria.
2. Complete the wiring and cables megger test before starting any termination for the system components.
3. A confirmation letter from all suppliers that the control panels and installation of their supplied mechanical systems are completed, complying with the manufacturer recommendations and philosophy of operation, and are ready for testing and operation.
4. Visual inspection and check for the installations to ensure everything is completed properly for the following:
    a) The panels are in good condition and without any damage.
    b) Ratings and settings of the breakers, contactors, timers, VFD starters, relays, fuses Etc. are set according to the approved workshop drawings and the approved mechanical load.
    c) All loops and panel components are identified and labelled using engraved labels and ferrules.
    d) All power and control installations are completed, and the procured material are provided according to the approved material submittal, the approved workshop and coordination drawings.
    e) Mechanical check for the inside material in the panels to ensure no damage.
5. Function test:
    a) All breakers are functioning mechanically in the correct way for (on/off/ and trip).
    b) A disconnecting switch or emergency push button is provided for each pump near the pump room door as applicable.
    c) All indication lamps are working (on /off and fault).
    d) Anti-condensation heater is working properly.
6. Final approval for the installations is subject to successful testing and commissioning. And approval of the mechanical engineer.

Operation and maintenance manual for control panels (mechanical installations).

1. Description of the project system.

2. How to use the manual.
3. Troubleshooting guide (for emergency breakdown).
4. Safety instruction at work.
5. Start-up and shutdown procedure.
6. Original catalogue with highlights on the material used.
7. As built, the drawing includes a schematic riser diagram/block diagram.
8. List of spare parts.
9. Procedure for routine maintenance for monthly/1 year/5 years.
10. Authority NOC.
11. Warranty certificate for 2 years (as per project specifications).
12. Testing and commissioning certificates/reports.
13. Emergency information.
14. Manufacturer/supplier information and contact details.
15. Overview and method of operation (sequence of operation).
16. Training for the security, operation, and maintenance team.
17. All guidelines for using all devices/controllers, etc.
18. Programming, settings, configurations, and the system software (if applicable) for the system.

Chapter 5

Containment

This chapter describes the wire and cable containment that includes the following:

1. UPVC ducts, PVC and Metallic G.I. conduits and accessories.
2. G.I. Cable Tray, Ladder, Basket, and G.I. Trunking

UPVC ducts, PVC and Metallic G.I. conduits and accessories

PVC containments are required in the early stage of work and need to be used in concealed installations in the beams, columns, slabs, rafts, and ceiling. Hence, they should be approved and procured early. The consultant should follow the contractor to submit the material submittals for these components for early approval and, hence, early procurement for the early start of the work.

G.I. conduits are required for ceiling and wall mounted installations; hence they will be installed after final finishing and painting of the room.

Both PVC and G.I. material should be procured with the required accessories to enable flexibility in installation and usually a sample board is submitted for consultant approval.

Approval of material submittal for PVC and metallic G.I. conduits

1. The consultant should instruct the contractor to obtain the consultant's approval for the used materials by submitting the material submittal. The following points are required to be considered for PVC and G.I. containment in the submittal:
    a. The manufacturer or brand is listed in the consultant's approved manufacturer list.
    b. Compliance Statement is provided as per project specifications.
    c. The original catalogue is included in the submittal. (Highlighting all technical points and requirements in the specifications).
    d. Submission of Samples.
    e. A list for all accessories to be used in the project. It should be included as part of the submittal like expansion coupling, bend, adapter, circular box, risers, Tees, etc....
    f. All bends for conduits above 25mm should be factory-made.
    g. Sizes of all conduits should be as per the approved load schedules for distribution boards.
    h. The submittal should contain all sizes required in the work, any conduit with size not shown in the submittal should be submitted separately.

❖ For PVC conduits and fittings:

PVC conduit to be used only for concealed application in front of house i.e., ceiling, wall, or slab. Except when used above false ceiling. (If the project using LS0H wires, then G.I. conduits should be used instead of PVC over false ceiling to fulfill the zero-halogen specification in the project).

- **For GI conduits and fittings:**

    a) G.I. conduit to be used only for surface application i.e., ceiling, (wall or slab) in service areas or back of house.

    b) For bends, a suitable bending machine is to be used with an experienced technician for correct bending.

2. The contractor should ensure that all materials arriving at the site have passed MIR (material inspection request) and that the material is complying with the consultant's approval.
3. The main contractor will ensure proper material protection inside the store and after installation at the site. (PVC conduits shouldn't be exposed to sun rays in the store; otherwise, they will lose flexibility and easily break.
4. Contractor must ensure that the correct revision of the approved workshop/ coordination drawings is being used. He should note that any clouded areas on the drawings are subject to revision, hence work should not proceed in these areas until an updated drawing has been issued and approved.
5. The contractor must ensure that all safety requirements are fulfilled, including suitable accessibility to the site. The site should always be cleaned and dried in accordance with the project safety plan.
6. The contractor should follow the approved method statement of installation and fixing the material to get the best quality of workmanship.
7. The contractor should submit the breakdown percentages for material installation for billing purposes. The Level of breakdown is to be agreed upon by all parties.

Notes for Installation of PVC conduits.

The consultant should instruct the contractor to start the installation of PVC containment on the site. The following activities should be checked and confirmed during and after the installation of the PVC Conduit in the slab:

1. Method statement for installation and fixing.
2. The type of concrete slab should be known (normal slab, post-tension slab, and readymade slab). Hence, conduit installation should match the used slab.
3. All outlets on the ceiling and electro-mechanical equipment should be marked on the slab shutter before 1st layer of steel is laid for easy guidance to locate the correct location of outlets or switches for Water Heaters, Fan Coils Units, Distribution boards, lighting point, 13A Switched Socket outlets, Spur outlet etc.
4. In case there is no false ceiling inside the room, the location of the light points should be coordinated with all other services like (fire alarm points, emergency light points, closed circuit television, speaker … etc.). light fittings to be located as (x – 2x – x) dimensions. also to be coordinated with bulkhead (if it exists) or internal decoration. **See the details in 5-1.**

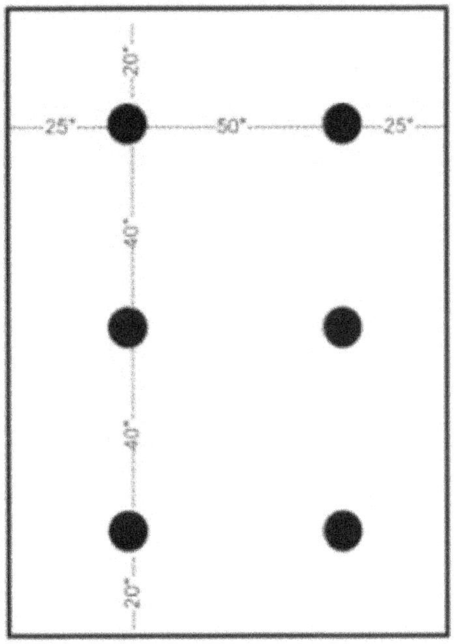

5-1 PVC conduits distributed in slab showing x-2x-x light distribution in the ceiling.

5. All conduits in concrete slab should be installed after the 1st layer of steel is laid, conduits installation directly on the shutter is not allowed. **See the details in 5-2.**

5-2 Pipes in slab between steel layers and in wall inside the wall

6. Slab conduits should be far away from any wet areas (like kitchen, bath water tanks, etc.) to ensure no water penetration to the conduits in future.
7. All conduits should be supported by a steel layer to fix them in place against concrete pouring.
8. The minimum spacing between two (2) conduits inside the concrete slab should be not less than 2cm to allow concrete to encase the whole conduit. **See photo 5-3**

9. An end cap is to be provided for all conduits after installation to keep the conduit clean and to ensure no dirt blocking it. (Water, dust, concrete, etc.).

5-3 Enough spacing between PVC conduits to be provided to enable concrete to encase the whole conduit and protect it.

10. All PVC boxes in concrete slab to be fixed by nail before casting to prevent movement when pouring the concrete. Nails should be removed directly after completion of casting and shutter removal, otherwise, it may change the color of the plaster and painting.
11. Where conduits cross expansion joints, a PVC expansion coupler must be provided at the position of the expansion joint.
12. All PVC conduit joints should be properly sealed with an approved sealant.
13. Check all draw boxes and junction boxes are located according to the Workshop DRG in accessable locations.
14. Dropped conduits to switches and D.B.s are to be arranged according to the approved drawing, always put spare conduits to replace the damaged once for future usage especially in recessed D.B.s in walls.
15. Check sleeves and openings in beams for continuous and straight installation are located as per approved drawings.
16. Ensure that the conduits related to the garbage chute riser are G.I. conduits.
17. The drop conduits of switches must be routed vertically without curves or angles as much as possible. All cuttings in walls related to switch/socket outlet should done using a grinder to ensure smooth cutting. See photo5-4.
18. All cuttings in concrete walls, slabs, columns, and beams should be approved by the Resident Engineer.
19. The depth of PVC conduit installed inside the wall and connected to the junction box shall be studied with the senior Civil Engineer to determine the thickness of plaster over the PVC conduit and to avoid future plaster cracks.

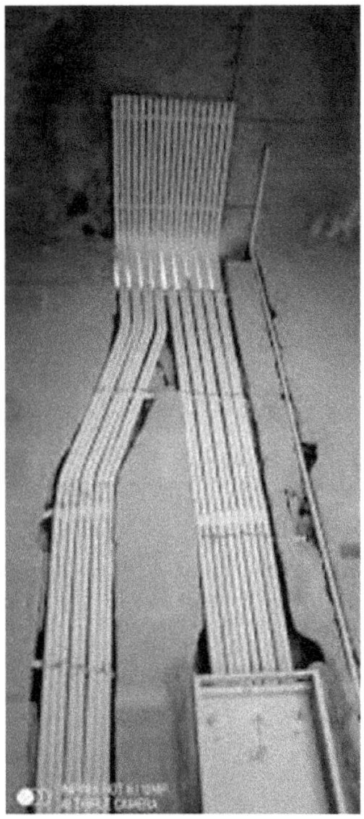

5-4 Concealed installation of PVC conduits requires cutting the wall properly with a grinder with suitable depth to allow at least 2 cm of plaster to be used over the conduits. Less plaster thickness may lead to future cracks. Work is to be coordinated with civil and architectural engineers.

20. All conduits connected to power socket should be from bottom, while lighting conduits should be connected to boxes for wall mounted lights form top.
21. The conduit for all external boxes and in the roof (for socket or switch or any junction box) shall be connected to the box from the top or from the side; connection from the bottom is prohibited to avoid water dropping inside the pipe if the gasket is damaged.
22. A plastic cap for all conduits is to be provided before concrete casting and before the starting of plaster.
23. For all D.B.s installed in areas with false ceiling, provide at least two extra conduits 2x20mm from D.B. to above false ceiling for probable future requirement.
24. All exposed floor conduits (not installed) will require protection by concrete casing for protection until tiles are fixed. It is advisable that such exposed floor conduits be installed late, just before tile work, to keep them from damage.
25. Worker (Electrician) on site should have the approved workshop, coordination drawings and installation details drawings and any other required details for the installation in the area.
26. For the water tank, all control and power supply conduits should be installed outside on the wall above the water level (Elevation Drawings and sketches are required for approval).
27. Conduits and fittings in the structure should be firmly tied and fixed without deformation to ensure rigidity to withstand vibration during concrete pouring. The electrical foreman should be on the slab when pouring concrete to keep concrete pump labors away from electrical conduits so as not to pour the concrete on PVC conduits and damage them. **See photo 5-5.**

Fig 5-5 when pouring concrete, the concrete pipe should not be directly over the PVC conduits for electrical installation to protect them from damages.

28. In case a Junction Box, PVC Box or G.I. Box are not available, Thermocol should be provided between the plywood shutter and reinforcement steel to provide room for switch box installation in future. **See details in 5-6**

5-6 Thermocol in the wall before shuttering to provide future space for the G.I. box.

29. Junction boxes are to be fastened securely from movement to ensure they will stay clean from the poured concrete.

30. All PVC conduits connected to boxes or panels should be connected by a suitable adaptor. **See the details in 5-7.**

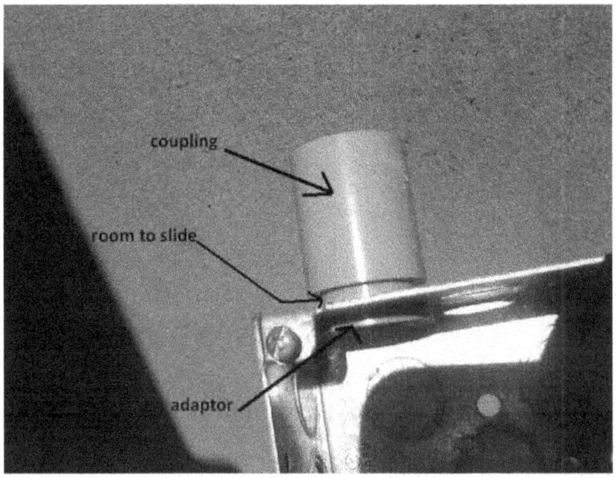

5-7 PVC conduit connected to the panel or G.I. box by Adaptor.

31. Any bend above the size of 32mm should be factory made (no site fabrication shall be allowed except for critical cases) site made bends should be Approved by the consultant in advance.
32. All walls should be marked on the slab by the main contractor to secure the dropped conduits to pass in the middle of the wall and to avoid any drop conduit from penetration outside the wall.
33. Ensure that all conduits connected to the distribution panel are passing in the middle of the wall in which the D.B. is installed (recess installation). See the details in in photo 5-2.
34. The maximum runs between two (2) outlets should not exceed 25 meters for straight runs and 10 meters for runs with bends to allow easy wiring. Pulling boxes should be provided every 20m or wherever necessary and the same to be highlighted in the drawing with Engineer's approval (pulling boxes should be in accessible area).
35. For easy wiring and future rewiring, a conduit's bend of less than 90° is not allowed.
36. Pendant conduits through main beams should be avoided. Conduits to be placed in the slab passing through the beams, they should be uniformly distributed. The outside diameter should not exceed one third of slab thickness.
37. Conduit in slabs should be placed in parallel to the main re-enforcement steel slab without touching it. Conduit arranged in slab (ceiling/ground) shall be distributed in between two layers of slab steel. Conduits touching slab shutter should be avoided. Conduit installation below the screed is not allowed.
38. Conduit should be sized in accordance with IEE or local authority but should not be less than 20 mm for lighting and 25mm at least for power and low current systems following the approved Load Schedule. For a 2-way switch, 25 mm2 is to be used.
39. Conduits shall be distributed in slabs in straight lines paralleled together as much as possible.

40. While two or more conduits are running in the same route, they should be placed at a distance at least twice the diameter size of the conduits. **See the details in 5-8.**
41. Conduits should be tied with steel wire at every 50cm and next to every junction box and coupler for proper fixing.
42. If conduits are required to pass through beam or concrete, or if spares to be provided for future, then sleeve pipe should be provided before concrete or conduits to be provide for future joining with the future installed conduits). No chasing is allowed in concrete.

5-8 Parallel installation of conduits with uniform distance between conduits. Conduits are fixed by saddles.

43. In the kitchen, toilets and other wet areas, conduits are not allowed to pass in these areas or below the water pipe/drainpipe/ valve, etc.
44. All openings in the slab/ drop beam are to be checked and marked on all drawings (power, lighting, etc.).
45. Where conduit cross expansion joint, expansion coupler must be provided at the position of the expansion joint.

Notes to be taken into consideration for the installation of G.I. conduits.

When using G.I. containment in the site in the installation stage. The following activities should be checked during and after the installation of G.I. Conduit. The contractor should be notified to take care of the following:

1. Method statement for G.I. conduits installation and fixing.

2. Complete metallic conduit system shall be used in accordance with the specification and approved shop drawings complying with British standard (BS-4568).
3. Conduits and fittings shall be manufactured especially for electric wiring purposes.
4. Conduits shall not be less than 20mm in diameter size and shall be complete with all necessary threaded fittings, coupling and connecting devices.
5. All conduits and fittings shall be free from rust or other defect and shall be properly stored in covered racking to avoid mechanical damages and weather effect.
6. Exposed conduit shall be installed parallel or right-angled to the direction of walls and ceiling beams; angled installations are not accepted. **See the details in 5-8**.
7. The parallel runs of the conduit shall be spaced uniformly. See the details in 5-6
8. Conduit shall be secured by means of space bar saddles. Spaced note more than 1.5 – 2.0 and 2.4 Cm for 20mm, 25mm and 32mm size respectively. **See the details in 5-9**.

5-9 G.I. conduits installed with parallel paths and fixed on the wall by saddles.

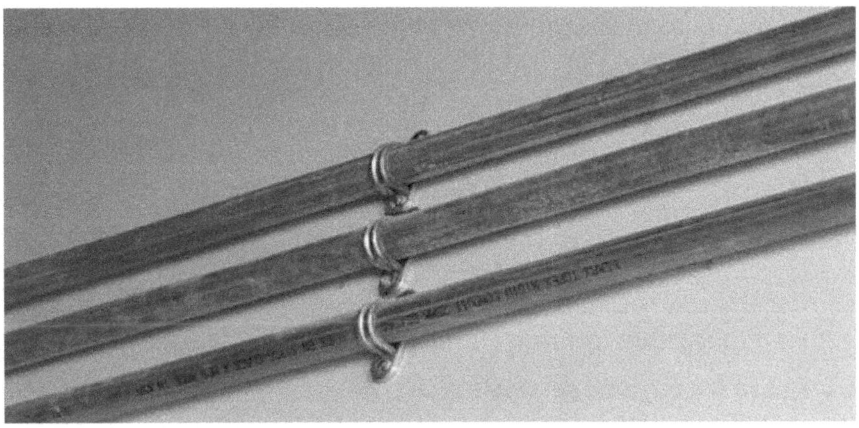

9. The inside surface and the ends of conduits and fittings should be cut smoothly, filed, and cleaned. They should be free from burrs and other defects; all cuts to be galvanized and painted.
10. All threads shall be cleaned and free from rust.
11. The installed conduit outlet should be closed temporarily by suitable plastic plugs for protection before wiring. **See the details in 5-10.**

5-10 temporary caps on the conduit to keep them clean.

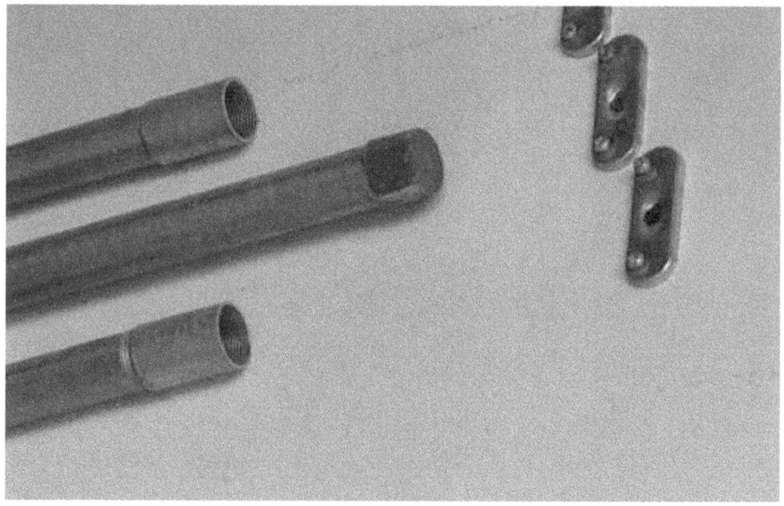

12. In damp and wet location, the entire conduiting system including boxes, fittings, panel boards etc. shall be mounted so that there is at least 6mm air space between it and wall supporting surface and shall be corrosion proof.
13. Flexible metallic conduit shall be used for connection to the motor or to other equipment to allow vibration or adjustment. **See the details in 5-11.**

5-11 Flexible conduits are used for the final connection to the electrical components.

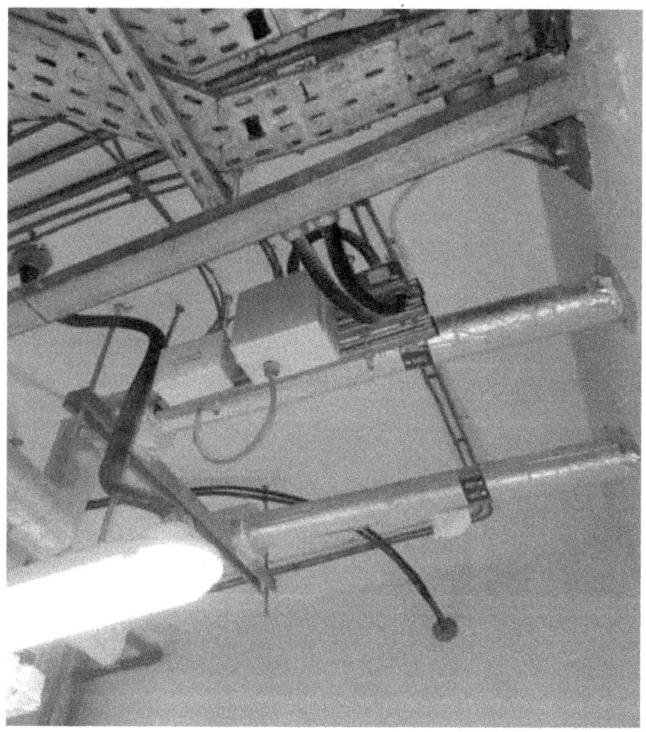

14. Draw wire shall be left in every conduit for installation of wires and cables and slack shall not be less than 30cms at each of the draw wire.
15. It is prohibited to use PVC boxes in G.I. conduit routes as it will cut the earthing continuity. Earthing continuity is to be ensured in the route.

16. All G.I. conduits should be connected to their related G.I. boxes by a suitable adaptor. **See the details in 5-12.**

5-12 G.I. conduit is connected properly to the G.I. box.

UPVC sleeves (ducts) for Authority service entry

For service entry to the project, the contractor should follow the authority regulations for the required quantity and size of the used sleeves.

For power cables, it may be a rectangular opening or 50 mm PVC sleeves (2 or more). For telephone, 2 or 3 sleeves may be required with a suitable size (50mm). Always refer to the latest regulations for the authority for the requirements. **See photos 5-13.**

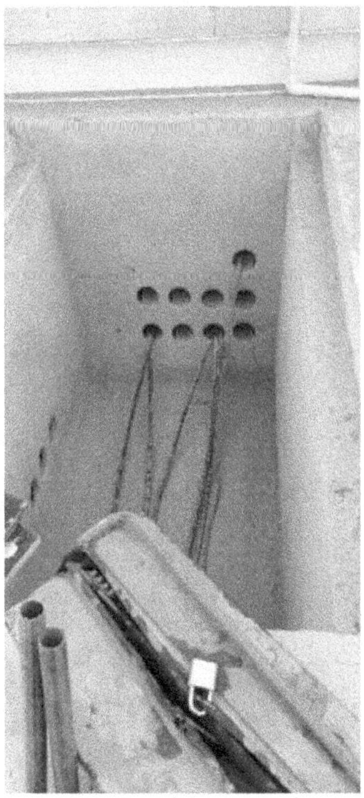

5-13 Entry sleeves for power cables to the building.

Special case for the sleeves in the sheer wall in the basement. These sleeves can be fixed in the shutter with bell mouth to the inner side of the building, hence when shutter is removed, the sleeves can be extended using the bell mouth for connection to the next pipe.

Always, ducts or sleeves should be inclined towards outside the entry of the building to prevent water penetration inside the building. Water seals can be used to prevent water from entering the building. **See photo 5-14**

5-14 cable sleeves closed by water seal to prevent water from entering the building.

G.I. Cable Tray, Ladder, Basket, and G.I. Trunking.

Approval of material submission for G.I. cable trays, ladders, baskets, and G.I. trunking.

The consultant should instruct the contractor to provide the material submittal for G.I. containment (Trays, ladders, Basket, and trunking) for the consultant's approval before starting installation work. The following points to be considered for the material submission and approval:

1. The contractor should obtain the consultant's approval for the used materials, taking into consideration the following points:
    a) The manufacturer is listed in the consultant's approved manufacturer list.
    b) Compliance Statement is provided as per project specifications.
    c) The original catalogue is included in the submittal. (Highlight the used components with part no and all technical points and requirements according to the specifications.)
    d) Submission of Samples.
    e) Equipotential bonding is to be provided for trays and trunking.
    f) Covers for all trunkings, external cable trays and ladders are to be provided.
    g) Before ordering the material approved workshop drawings should show the size of the cable trays or ladders to be used. Trunking sizing will be based on numbers of wires inside according to the shop drawings taking into consideration the recommended space factor. Approved shop drawing should show the trays/ trunking routing.
    h) Heavy-duty cable trays shall be with return flanges.
    i) All required accessories, like bends, couplers, etc., are to be factory-made.
2. The Main Contractor should ensure complete protection for trays, trunking, Baskets, and ladders inside the store and after installation.
3. Support calculations for trays, trunking, Baskets, and ladders should be completed and submitted prior to installation considering the following points.
    a) The total cable load on the cable tray per meter.
    b) Load carrying capacity of fixing bolts.
    c) Actual size of the fixing bolts
    d) Actual size of the suspended metal rods.
    e) All the above should be in accordance with acceptable mechanical stresses and sag.

4. Contractor to ensure that materials arrived at the site have passed MIR (material inspection request) and being approved by the Consultant.
5. Valid supplier trade license with expiry date, and ministry of trade certification as a legal supplier of the material in the submittal.

Notes for Installation of Cable Trays, Ladders, Baskets, and Trunking.

The consultant should notify the contractor about the following:

1. The installation Method Statement is to be submitted and approved before starting installation work.
2. The contractor should ensure the best quality of work based on the approved method statement for installation and fixing of the trays, ladders, baskets, and trunking.
3. Contractor should ensure that only the latest approved revision of workshop/ coordination drawings is being used. Clouded area on the drawing is subject to be revised and work should not proceed in that area until an updated drawing is being issued.
4. The contractor must ensure that all safety requirements are fulfilled, including suitable accessibility to the site. The site should always be cleaned and dried in accordance with the project safety plan.
5. The contractor should ensure the received material in the site is being inspected by the consultant and approved, MIR is signed.
6. The contractor should ensure full protection for the material inside the store and after installation.
7. The installations should be coordinated with mechanical work (ducts, water pipes) and any other services in the same area based on the approved coordination drawings.
8. Vertical support positions are to be Marked from the grid line and to be coordinated with other services. Location will be based on the approved coordination drawings (spacing between supports). Threaded rods and supports should not be used to support more than one service, this will enable easy future maintenance without affecting other services.
9. All openings in the walls/ slabs, etc. matching the sizes and locations of cable trays/ ducts/ trunking and baskets etc.
10. The routing of cables should be easily accessed for installation and maintenance.
11. Heavy-duty cable trays should be used for power cables and light-duty cable trays for low-current cables.
12. Cable trays and their accessories are off hot dip galvanized finish.
13. The suspended supports are at regular intervals of a minimum of 1200mm and not more than 150mm from all bends, tees, intersections, and risers.

14. All trays and trunking joints are positioned as close as practicable to the cable tray and trunking supports.
15. A minimum clear space of 25mm shall be kept behind all installed runs of cable tray and from the wall.
16. 25% of the cable tray size should be considered for future uses if the cables are arranged in a single layer on the cable tray.
17. All cables in the cable trays are directly supported or tightened by cable straps or cable ties to maintain the form of a sleek and uncluttered. Vertical runs off the cables are to be supported with cable cleat or saddles.
18. All field cuts on parts or components are painted with anti-corrosion paint, Galva fiord.
19. Ensure tightness of all screws, nuts, and washers.
20. All adjacent lengths of the tray or trunking are joined using joining couplers.
21. All fittings for change of direction or elevation should be selected and factory-made. See 5-15
22. Proper earth continuity and bonding are connected to the main building earthing system as required.
23. Ensure correct identification is provided. **See 5-15**
24. Ensure provision of fire-rating materials between all openings for floors or between rooms that are related to different fire zones.

5-15 Cable trays and trunking with factory-made fittings and accessories. Trays identification is provided on trays.

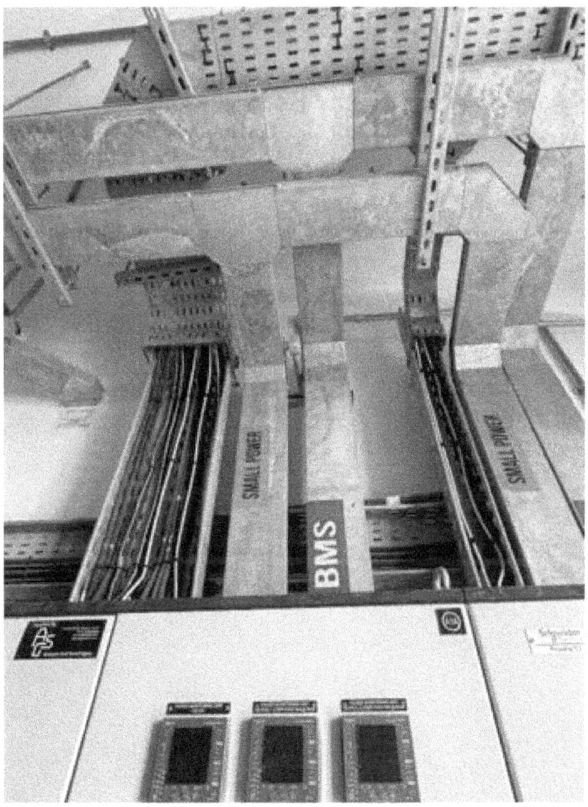

25. Distance between all H.T. cables and other services to be 45cm (no service should be below or above cable tray), as per authority requirement.
26. All floor cable trays and ceiling cable trays crossing below the wet pipe should be covered by a G.I. (galvanized) steel sheet.
27. Multilayer trays should allow proper cable pulling and calculations to be submitted for approval.
28. Breakdown for all items related to the materials shall be proposed by the Contractor for billing purpose, the percentage of the installed components shall be agreed by all parties.

    Note: Usually, baskets are used for low current systems as the openings in the basket doesn't allow easy cable bending for diverted cables. **See 5-16 for details.**

    For power cables, Ladders can be used in the electrical room as they allow more flexibility for cable diversion through the ladder steps.

    Trunking and trays need tees and bends, which need more space and more difficulty in restricted areas.

Operation and maintenance manual for Trays, Ladders, Trunking and baskets.

1. Description of the project system.
2. How to use the manual.
3. Safety instruction at work.
4. Original catalogue with highlights on the material used.
5. List of spare parts.
6. Procedure for routine maintenance for monthly/1 year/5 years.
7. Warranty certificate for 2 years (as per project specifications).
8. Emergency information.
9. Manufacturer/supplier information and contact details.

5-16 Basket is used for small cables and low-current cable installation.

Chapter 6

Installations (Part 1)

This chapter describes the installations of the following:

1. Junction Boxes and Back Boxes.
2. Wires and Cables.
3. Bus Bar Duct.
4. Switches and Sockets Outlets.

Junction Boxes / Back Boxes

Approval of material submittal for back boxes

1. The consultant should follow the contractor to submit the material submittal to get the consultant's approval for the used materials, taking into consideration the following points:
    a. The manufacturer is listed in the consultant's approved manufacturer list.
    b. A compliance Statement is provided based on project specifications.
    c. Original catalogues are included in the submittal. (Highlighting all technical points, part no for the used components, requirements in the specifications). Separate submittal for all other junction boxes related to (Tel, T.V., Intercom, floor box, etc...) to be submitted.
    d. Earth Terminal is provided.
    e. Submission of Samples.
    f. Installation Method statement.
    g. Purpose and location of usage (For switch, socket outlets, etc.).
    h. Phase barriers are provided for switch panel boxes.
    i. Adjustable lugs are included.
    j. Minimum depth of 35mm for switches, sockets, etc....
    k. Minimum depth of 50mm for cooker outlet.
2. The Main Contractor should ensure complete protection for boxes inside the store and after installation.

Points to be considered before starting to mark, fix, and install boxes.

1. Contractor must ensure that the correct revision of the approved workshop/ coordination drawings is being used. He should note that any clouded areas on the drawings are subject to revision, hence work should not proceed in these areas until an updated drawing has been issued and approved.
2. Contractor should ensure the best quality of work based on usage of the approved method statement for installation and fixing the materials.
3. The contractor must ensure that all safety requirements are fulfilled, including suitable accessibility to the site. The site should always be cleaned and dried in accordance with the project safety plan.
4. The datum line information to be provided by the main contractor with the final FFL (finish floor level) for correct marking.

5. Mounting heights are to be marked on shop drawings or referenced to the approved mounting height schedule.

6. Marking of the boxes in the wall to be done according to the approved shop drawings, furniture layout, coordination drawing for all services areas (H.V. and transformer room, telephone rooms etc.).

7. All used material should be checked to be as per approved submittal.

8. The contractor will ensure that all materials arriving at the site have passed MIR (material inspection request) and are approved by the consultant.

0. The boxes should be of adjustable type to allow adjustment of wiring accessories, contain copper/brass earth terminal and be of minimum 35mm depth.

10. Boxes for different services are to be mounted at one level as approved in the schedule of heights with uniform distances between boxes of same size of wiring accessories.

11. Wall cutting for the builder's work should allow extra openings and cuttings to allow the cement mixture to be inserted properly to fill the gaps.

12. Wall cutting and chasing should be in straight lines to the ceiling.

13. Use suitable tools to fix the back box correctly.

14. Ensure the door opening direction and fix the boxes, accordingly, ensure coordination with architecture for the thickness of door frame or windows.

15. Extra care should be taken when fixing boxes inside concrete structure columns or RCC (reinforced concrete) walls so as not to move while pouring the concrete; thermocol can be used to ensure enough room for future fixing of the boxes. **See 6-1**

6-1 Thermocol is fixed to provide future room for G.I. boxes in the steel reinforcement; the thermocol should be fixed properly so as not to move after fixing the shutter or pouring the concrete

16. Back-to-back box installation is not allowed unless approved by the consultant.
17. Conduit's installation can be done after back boxes and panel boxes installation is approved by the consultant, all conduits' entries to be closed and boxes to be covered for protection from dirt and concrete.
18. The connections between the conduit and the PVC box, G.I. box, and D.B. box should be through a suitable PVC/ GI adapter, depending on the used container.
19. Check and ensure that the conduits are connected correctly to their respective D.B.
20. The distribution panel should be installed on a wall with a thickness of not less than 20 cm (coordination with civil or main contractor to ensure the wall is at least 20 cm thick for D.B. installation). The D.B. should be installed to match the final plaster level. All boxes installed in walls are fully aligned and flushed with the plaster level after the plastering of the wall is done (the thickness of the plaster should be marked on the wall by the civil contractor).
21. For cooker outlet (switch and flexible outlet), The depth of G.I. box for cooker outlets should be 50mm to facilitate cooker wiring connection (6 mm or 10mm size).
22. Recessed outlet boxes for surface-mounted devices of lighting fixtures must be totally concealed behind the device or fixture (location to be coordinated with fixture base).

23. Ensure using a suitable back box for the outlet of the system (Intercom, Access control, Sonder, thermostat ..etc).
24. The depth of outlet boxes should be selected to suit the numbers and sizes of conduits and conductors.
25. Check that the edges around the boxes are adjusted such that they fix the walls properly. This is to be done by the Main Contractor during plastering when the plaster is still wet. **See 6-2**

6-2 The plaster around the switches' boxes should be repaired and done properly before the final finishing of the paint.

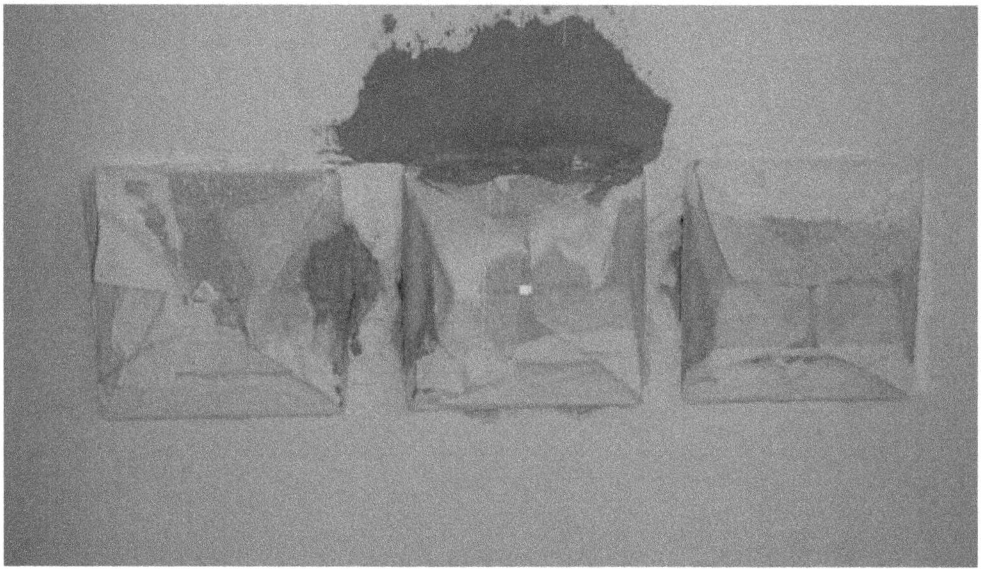

26. Ensure G.I./boxes, Junction Boxes, and D.B. boxes are not installed below or in front of the water pipe/drainpipe/valve, inside the kitchen, bathroom, washing machine or the cleaning room, wet area, or staircase.
27. If switch box contains more than one phase (group switching), approved switch boxes with phase barriers should be used and labeled to indicate that 400 volts exist at the box. All circuit wires should be color identified as specified in the authority regulations.
28. No switches or outlets controlling lighting circuits or water heater, cooker switch, oven switch, dishwasher switch, washing machine switch, ... etc., should be installed in the bathroom. In the kitchen and in situations where water is regularly used, no switching should be mounted within 2m of any water tap, wash basin, or sink. If this is not possible, ceiling-mounted insulated cord-operated switches shall be used. (Refer to authority regulation).
29. Use a suitable W/P (weatherproof) back box for all W/P outlets.
30. Boxes in the roof should be installed that the conduit is connected to the box from top or upper side to prevent water leakage to the pipe in case of water penetration to the box.
31. Breakdown for all items related to the materials shall be proposed by the Contractor for billing purpose, the percentage of the installed components shall be agreed by all parties.

Wires and cables

Approval of material submittal for wires and cables

1. The consultant should instruct the contractor to obtain the consultant's approval for the used wires and cables, including the following points:
    a. The manufacturer's brand is listed in the consultant's approved manufacturer's list.
    b. A compliance Statement is provided based on project specifications.
    c. The original catalogue is included in the submittal. (Highlighting all technical points, used parts with part no, and requirements in the specifications).
    d. Separate submittal for all accessories related to wires and cables (cable ladder, cable tray, cable gland, cable identification, etc.).
    e. Submission of Samples.
    f. A table showing all cables, types of insulation (PVC/LSZH/XLPE, heat resistant, fire resistant, Armored, or non-armoured), sizes, lengths, and related voltage drops.
2. Valid supplier trade license with expiry date, and ministry of trade certification as a legal supplier of the material in the submittal.
3. The type of cable for the power supply of the fire alarm system and firefighting equipment should be as per Authority regulations.
4. Fire rated cables shall be considered for all life safety and firefighting equipment including but not limited to the followings to maintain safe working conditions during fire.
    a. Firefighting lifts
    b. Staircase and fire lift lobby pressurization fans
    c. Firefighting pumps (sprinkler and wet riser pumps)
    d. Electrically operated fire shutters/curtains
    e. Emergency central battery lighting equipment.
    f. Generator

Points to be considered before installation of low-voltage wires and cables.

1. The contractor must ensures that the latest correct revision of the approved workshop drawings is being used; if there is any clouded area on the drawings, then this is subject to revision, and work should not proceed in these areas until an updated drawing is issued and approved.

2. The contractor must use the approved method statement of installation of materials to get the best quality of workmanship.
3. Electricians should be skilled and aware of the correct materials handling and installation procedures to ensure the best quality of workmanship.
4. The contractor must ensure that all safety requirements are fulfilled, including suitable accessibility to the site. The site should always be cleaned and dried in accordance with the project safety plan.
5. Before ordering the cables, ensure the followings to procure the correct cables:
    a. Authority approval for all cables' sizes based on the schematic diagrams, connected loads, cable lengths ensuring the allowed drop voltage according to the authority regulations.
    b. Approval for all mechanical loads by the mechanical consultant.
    c. All cables are to be ordered and routed without joints.
6. Wiring in conduits, sizes, and number of wires should be referenced to the specifications, drawings, and load schedule.
7. All wires and cables should be tested for continuity and insulation test before installation.
8. Routing of cables and wires shouldn't affect any other low-current systems.
9. Cables and wire identification using engraved labels with cable ties and suitable wiring identification are to be provided. See cable and wiring identifications in 6-3
10. The fire resistance cables shall be presented in power schematics drawings in a similar way to the normal armoured cables, as shown in the examples below:
    a. Normal XPLE armoured cable (4x300mm2 XLPE ARM + ECC 1x150mm2).
    b. The fire-resistant cable (4x300mm2 MGT/XLPE/LSZF/SWA/LSZH + ECC 1x150mm2).
    c. Normal PVC armoured cable (4x6mm2 PVC ARM + ECC 1x6mm2).
    d. The fire-resistant cable (4x6mm2 MGT/PVC/LSZF/SWA/LSZH + ECC 1x6mm2).
    e. The normal single wires/cables installed in PVC conduit will be used as fire resistance cables if embedded / concealed in the concrete slab.

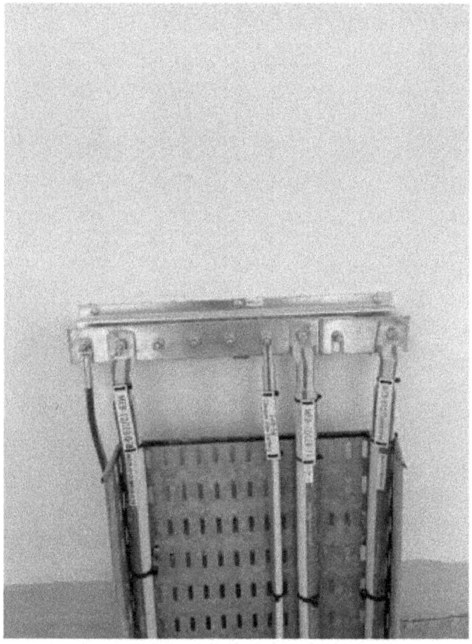

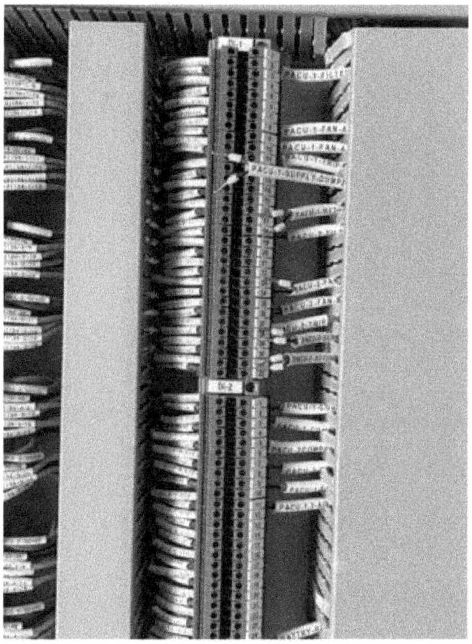

6-3 Cables and wiring identification

11. The cable legend abbreviations shall be included in the legend drawings under "Abbreviations", as shown in the following examples:
    a. MGT – Mica glass tape.
    b. LSZP – Low smoke zero fume.
    c. LSZH – Low smoke, zero halogen.
12. Additional information and abstracts related to cable specifications/ standards are shown below for guidance and information:
    a. Fire-resistant power cables, according to BS6837.
    b. Conductor: Plain annealed stranded copper according to BS6360 (CLASS 1).
    c. Fire barrier: MGT (Mica glass type).
    d. Insulation: XLPE (cross-linked polyethylene).
    e. Core identification: 2 cores: Brown and Blue, 3 cores: Brown, Black and Grey, 4 cores: Brown, Grey, Blue & Black. Multicore: White cores with black numerals.
    f. Bedding: LSZP (Low smoke, zero fumes).
    g. Armour/Protection: SWA (Steel wire armour).
    h. Sheath/Jacket: LSZH (Low smoke, zero halogens).
    i. Colour: Black.
    j. Voltage: 600/1000v.

   k. Operating temperature: Maximum 90 Degrees C, minimum bending -25 Degrees C.

13. Some of the applicable Standards:
    a. BS6387 Categories C, W & Z, IEC331, BS EN 61034:2005, BS6724, BS7846 & BS EN 50266:2001.
    b. FIRE RESISTANCE cables according to BS6387 Categories C, W & Z:
    c. IEC60331: 750 Degree C for 3 hours.
    d. Flame Propagation according to BS EN 50266 parts 1 & 3, IEC 60332-3, BS EN 60332:2004.
    e. Smoke Emission according to IEC 61034, BS EN 61034:2005.
    f. Corrosive Acid Gas according to B.S. 50267-2-1:1999, IEC 60754-1.
14. Before ordering the materials. The contractor must get approval from the Authority and consultant for the following:
    a. Workshop drawings (schematic diagram, layouts, details, and any drawing required by the authority).
    b. Mechanical equipment (actual mechanical load)
    c. A list for all sizes of all cables' lengths (cable joint is not allowed), size and voltage drop.
15. The Main Contractor should ensure protection for materials inside the store and after the installation of wire and cable.
16. Breakdown for all items related to the material shall be proposed by the Contractor for billing purposes, the same percentage shall be agreed by all parties.
17. The contractor is to be sure that all materials arriving at the site have passed MIR (material inspection request) and approved by the Consultant.

Installation of Low Voltage wires.

The consultant should notify the contractor to ensure the following points during the installation of Wires:

1. When pulling the wires, the wires should not be placed on the ground and pulled out for long distances as this may destroy the insulation.
2. Check the wires insulation by insulation tester to ensure that wires are healthy and free of damages during transportation or any miss-handling.
3. Arrange the wire bundles to be rolled easily during Pulling.
4. Make sure that all electrical conduits/pipes that wires will be pulled inside are clean and free from blockage/obstruction or sharp edges.

5. Ensure only one single-phase circuit per conduit is pulled.
6. Ensure approved load schedules are available with the Electrician on-site during the installation of the wires.
7. The circuit wires of individual categories, such as lighting, power, emergency, etc., are to be segregated with barriers in trunking runs or installed in separate conduits.
8. For residential premises supplied with a three-phase supply. Ensure light fittings, socket outlets, water heaters, cookers and other single-phase apparatus installed within one room are not connected to more than one phase as far as possible. If more than one phase cannot be avoided in supplying the apparatus in one room, a minimum distance of 2.0 meters is to be maintained between any two fittings/outlet/accessory or appliance with different phases (refer to Authority regulation).
9. Ensure earth conductors are the same size of live conductor (up to 16mm2), or according to the authority standards.
10. Ensure a slack of 150mm wires is left at both sides of the service box and junction box (if any) for easy termination. Wires and boxes to be protected from dirt.
11. Pulled wires between two points should be continuous without any joint. If the joint is necessary, it should be through junction/outlet boxes.
12. Check that the conductors at every outlet, junction box, and D.B. are well identified (line, earth, and neutral).
13. Check that each metal box is earthed with live Earth conductors and to be bonded with the device.
14. Check that the terminations of all wires at D.B.'s are done with pinpoints in lugs.
15. Check wire insulation by Megger before starting termination to discover damages in wire insulation (megger test).
16. Check correct wire and conduit sizing, number of wires, and ensure it is complying with the specifications, local standards, and approved load schedule.
17. Check that all pulled wires are the correct type and size as per the approved load schedule.
18. Check that all connections between the point and lighting fitting are heat resistance wire/cable.
19. Check that the colour of all wires (R, Y, and B) matches the approved load schedule.
20. Check that the contractor is using the correct type of wires and check the type of insulation of the wire used (PVC/heat resistance/MICC), etc.

General notes for Low Voltage Cables:

The consultant should notify the contractor to ensure the following points during the installation of cables:

1. The size, type, and length of the cables should be as per the approved drawings and submittal. The selected Make/Manufacturer should be from the Approved list.
2. Cables to be stored properly to protect them from damages. Cables to be cut from drums wisely so as not to waste a lot of cable cuts. (It is preferred to create scheduled for similar cables and assign the drums to cut a specified cables with specified length from each drum).
3. All temporary and unused cables (if they exist) above the suspended ceiling or under the ground should be removed prior to Consultant inspection.
4. Ensure approval for all related Drawings from concerned Authorities and the consultant before starting the installation. Latest stamped and approved drawings by the consultant should be provided at site for execution of work.
5. Engraved labels should be provided for all cables at both Ends, bends and entry of ducts (as mentioned in the project specifications). **See the details in 6-4.**

6-4 Engraved labels are provided for all cables for cable identification.

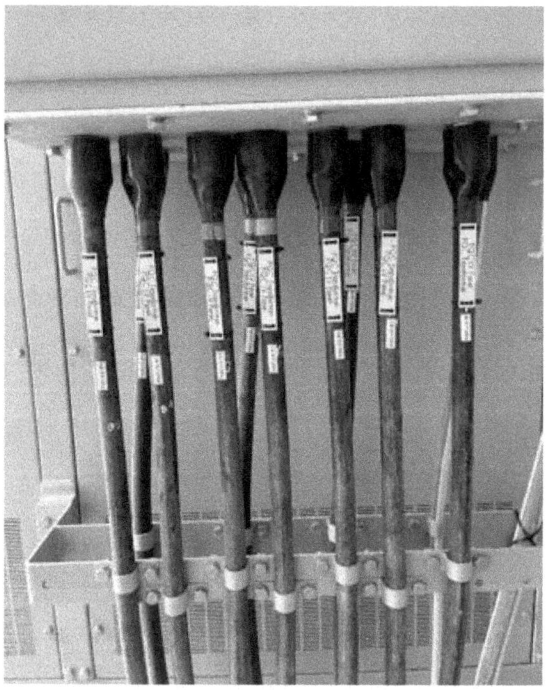

6. A damaged cable or cut pieces/ joints in the cable are totally prohibited. Joints shouldn't be used except for very long runs (more than the cable drum length) that require joints; the joints are to be replaced in places approved by the consultant and to be identified and shown in the as-built drawing.
7. Breakdown for all items related to the material shall be proposed by the contractor for billing purposes, the same percentage shall be agreed by all parties.

8. The contractor should ensure that all materials arriving at the site have passed MIR (material inspection request) and approved by the Consultant.

9. For all stages of the executed works, cables should be well protected during and after installation. (Cable pulling, Installation, termination, testing… etc.). Installation shouldn't proceed without cleaning the site and removing everything may damage the cables when cable pulling.

10. The followings should be ensured before the termination of the cable:

    a) All cable glands are used with the correct size and function (waterproof, armoured cables, explosion-proof. etc.).

    b) All cables are terminated with the correct size of lugs using a hydraulic crimping machine. See the details in 6-5

6-5 Cables are terminated to the bus bars with correct lug sizes, crimped with a Hydraulic crimping machine, and covered with sleeves with the same phase colour.

    c) The size of the earth conductor should be as per the standards and approved drawing, dressed uniformly along with armoured cable.

    d) A preliminary test for insulation of cables using a Megger tester should be done before and after installation of the cable.

    e) Any damage to outer insulation should be rectified using cable jackets as per authority regulations and manufacturer recommendations.

    f) Ensure that fireproofing sealants are used inside all Electrical rooms/service rooms for all incoming cables to avoid fire penetration from one area to another as per Authority regulation. **See the details in 6-6**

6-6 Fire proofing sealant is used to close all services openings in the wall between two fire zones.

General notes for installation of ARM/XLPE cable on cable tray/in floor trench/UPVC duct/ direct buried in ground etc.

1. Single-core armoured cables are connected between two panels.
    a. The gland must be of isolated/ insulated type.
    b. All AWA cables should be secured by Aluminum glands. Brass or Copper glands are not accepted. **See photos 6-7**
    c. Single-core cables should be fixed by brass glands; steel glands are used with 3-phase cables only. **See photos 6-7.**
    d. If there is single core cables to be connected, then the landing plate should be made of aluminium. Steel is not allowed due to eddy current generation.
    e. Single core cable should be laid on trays, trenches or underground installations in trefoil formation to reduce the electromagnetic field effect.
    f. Only one end of the Gland should be connected to earthing system to prevent earthing current circulation. **See the details in 6-8.**
2. Installation of cables on cable trays:
    a. Cables are to be properly saddled to cable tray using suitable cable ties or cable cleats.

b. Cables on tray to be in one layer, the spacing between two cables to be as per standard. (2D Minimum).

c. For loaded trays, tray loading calculation to be done for cable Trays to ensure that the tray can withstand the installed cables, trays should stay straight and free from sag due to cables' weight.

d. All cut edges of cable trays, threaded rods and U-channels should be painted with galvanizing paint to prevent rust.

e. Only factory-manufactured/fabricated standard bends and offsets should be used; site fabrication for accessories is not allowed.

6-7 Brass gland for single core cables and Aluminum gland for Aluminum cables.

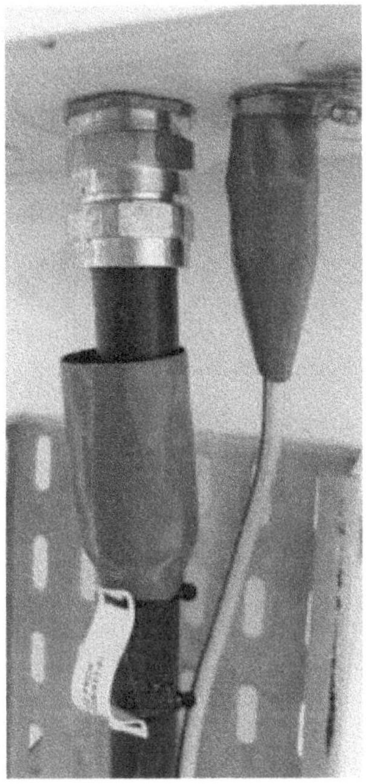

6-8 Gland earthing (for one side only)

3. Installation of cable in floor trench:
    a. The trench should be clean.
    b. Spacing between cables should be as per standard.
    c. The depth of the trench should be 900mm (Minimum). Or as in the approved shop drawing.
    d. Suitable sleeves to be provided from the trench to the wall for outgoing cables. (Coordination with civil work in the advanced stage).
4. Installation of cable inside UPVC duct:
    a. Both ends of the duct shall be sealed with Rigid Spray Foam to avoid entry of water/rodents.
    b. The depth of the duct should be 900mm (Minimum). Or as in the approved shop drawing.
    c. The minimum bending radius for the cable shall be within the standard limit. (In general, 8 X cable diameter for non-armoured cables, and 12X cable diameter for armoured cables).
5. Installation of the cable directly buried in the ground:
    a. De-watering should be done before laying cables. (If required).
    b. The depth of the installation should be 900mm (Minimum). Taking into consideration the level of the curbstone. (If any).
    c. The minimum spacing between the cables shall be twice the diameter of the cable.
    d. Ducts should be provided at the road crossings, between the Building and service manholes.
    e. Cable marker/ identification labels should be provided at both ends of the duct entry.
    f. The cable route should be furnished with tiles, warning tape and markers according to the project specifications and authority regulations.
    g. Backfilling should be done only after receiving clearances from all concerned parties. (Authority, Consultant, Sub-contractors).
    h. For all cables (LV, HV and control) cables, precast concrete tiles or PVC tiles should be provided above the cables with suitable marking (HV cables, LV cables , control cables) with warning tape and sand layers, according to the authorities' regulations.

Testing and Commissioning (Wire and Cable)

For wire and cable testing and commissioning, the consultant should instruct and follow the contractor to submit a T&C method statement with an approved checklist for test results records showing all tests to be done for the wires and cables with successful criteria based on the applicable standards and authority regulations.

Testing of final circuits in distribution boards

The contractor should perform the below tests and activities for the electrical circuits in distribution boards, the observations and results should be recorded in the check list for submission and record.

1. Visual inspection:
    a. Conduits, wires, trunking, light fittings, socket outlets, etc., connected to the final circuits of the distribution boards should be visually inspected for proper installation and against physical damages.
    b. Check the distribution board installations and internal wiring to be correctly finished mechanically and electrically, with correct identifications and labelling. Etc.
    c. All metallic parts should be earthed/ bonded.
2. Polarity check:
    a. Single pole switches are wired with the live conductor.
    b. All wiring accessories are wired correctly: the line wire to the Line point, the neutral wire to the N point, and the earth to the E point.
    c. Light fittings are wired correctly so that the switch will cut the line wire.
3. Continuity of protective conductors:

    For each final circuit, the continuity resistance of the protective conductor from beginning point to the main earth terminal should be measured with a D.C. Ohmmeter. The value recorded and verified should be within the permissible limit that not affecting the earthing pit resistance that is 1 ohm for earthing system and 10 ohms for lightening system.
4. Continuity for final circuit conductors:

    For each circuit in the D.B. (line, neutral and ground), check the continuity of each wire with the other after shorting in the source bus bars to ensure wiring continuity. The measured value should be acceptable and nearly zero (remember to remove the shorting wire).

Testing of Distribution cables (power and control cables).

The contractor should perform the below tests and activities for the power distribution cables, the observations and results should be recorded in the check list for submission and record.

a) Visual check:

   Check the laying of the power cable, landing, termination, support, labelling, identification, etc., and ensure all are done correctly.

b) Insulation resistance test for power cables:

With a 500V megger, measure the Insulation resistance between cores. Readings should be verified with the permissible limit from the standard or the catalogue.

With the same megger, insulation resistance between cores, armour, and earth terminals should be measured. Readings should be verified within permissible limits.

c) Phase rotation test:

After completion of tests (a) and (b) stated above, the relevant breaker should be switched on, and the phase rotation at the downstream point should be checked to ensure a similar phase sequence in all outgoings (i.e. RYB).

d) Performance test:

Upon verification of the correct phase rotation to the load, switch on the isolator (if any) and run the equipment to verify the complete performance of the connected equipment minimum for 24 hrs. to check for any abnormal temperature rise in the cables, terminals, switches, and protective devices etc.

e) Insulation resistance for wiring:

Insulation resistance between phase, neutral and earth to be measured with a 500 volts megger. (During this test, either all lamps and connected loads and apparatus to be removed from the circuitry with all switches closed, or all local controlling switched to be kept open and lamps and connected equipment can be kept in the circuit), values to be verified with the permissible limits.

Note: The main supply can be switched on to the final circuit to enable further tests that require power supply to be carried out.

f) Earth fault loop impedance test:

An earth fault loop impedance test is to be done for each final circuit with an impedance tester. (RCCB to be passed during the test) and the reading is to be verified within permissible limits.

g) RCCB test:

All RCCB (ELCB) protected branches in the D.B. are to be checked by means of an RCCB tester with a minimum of 3 different current settings (above normal, below normal, and normal sensitivity current setting of ELCB). The trip time is to be recorded. The values are to be verified with the permissible limits. (20 mA for power sockets, 100 mA for lighting circuits).

h) Circuit performance:

After conducting the above-mentioned tests with satisfactory results, an operational test should be conducted to verify the performance of the circuit. These tests are as follows:

a. Switch on the MCB in the D.B.

b. For the lighting circuit, operate all switches connected in the circuit to ensure proper control of the desired light fixture according to the approved switching performance.

c. For ring/radial circuits, test all sockets with a test lamp to ensure power availability.

d. For any other fixed equipment, operate the relevant switch and observe the functioning of the equipment.

i) Full load current:

Measure and record the current in all phases of the incoming cables in lighting D.B.'s by keeping all lights on.

Operation and maintenance manual for wires and cables

The consultant should instruct the contractor to submit the O&M manuals for the wires and cables that include the following:

1. Description of the project system.
2. Safety instruction at work.
3. An original catalogue highlighting the used material.
4. Warranty certificate for 2 years (as per project specification).
5. Testing and commissioning certificates/reports.
6. Emergency information.
7. Manufacturer/supplier information and contact details.

Bus Bar Duct

Approval for material submittal for the Bus Bar Duct.

The consultant should instruct the contractor to arrange periodical meetings between the Contractor's Electrical engineer, the supplier of the Bus Bar Duct and the supplier of Low Voltage Switchgear or Power Transformer (If the Transformer is connected to the Bus Bar Duct) to discuss and clarify the routing of the bus duct, connection to switchgear, Bus Duct rating, Authority requirement, material submittal requirements, work progress, and Isometric diagram, etc. The contractor should provide the followings for the consultant's approval:

1. The Bus Duct supplier is listed in the consultant's Approved Manufacturers List.
2. Description of the bus duct system and all related components.
3. Compliance statement to be submitted. The supplier should clearly show his compliance to the specifications, non-complied sections to be highlighted with justifications.
4. The submittal should include the type test certificates for the bus duct.
5. The submittal should include the original catalog for each component of the bus duct highlighting the technical points that exists in project specifications and highlighting the part number of the proposed components.
6. Details of Bus duct connection through the flange to the MDB or power transformer (based on the situation in the project) according to the approved workshop and coordination drawings and site conditions. (Transformer submittal and MDB submittal should show the provision for the busduct connection).
7. Ingress protection (I.P.) rating for the bus duct should be suitable for the location of the installation and should be according to the approved coordination drawings.
8. Fault level for the bus duct should be according to the authority requirements, (50KA for 2500A Bus duct, and 65KA for 3200A) or should be according to the approved design calculations. Fault duty for tap off unit MCCB should be 50KA.
9. The bus duct should be rated to 50C Celsius.
10. Busduct tap-off unit to be of the plug-in type.
11. Expansion joint to be provided where required.
12. The bus bar should be desisgned and procured with 5 plates (RYBN+E) or 4 plates (RYBN) according to the authority recommendations or project specifications. If RYBN is procured then a suitable sized earthing cable should be installed parallel with the bus duct.

13. Engraved labelling identification should be provided.
14. Coordination with switchgear or Transformer supplier for the following:
    a. Connection between the bus duct from both sides (MDB, SMDB, Transformer).
    b. The current setting for the ACB in the MDB should be suitable for the bus duct rating.
    c. Connection between the bus duct and the SMDB (if applicable)
    d. ACB in MDB and MCCB in the tap-off should be from the same manufacturer; the tap off should be plugged in.
    e. Discrimination and short current calculations between the MDB, SMDB, D.B., bus duct and tap off.
15. Coordinate with the civil and architectural engineers for the finish floor level and ceiling level for each floor, location, and dimensions of openings in the electrical room for the bus duct riser.
16. Coordination with mechanical engineer for the followings:
    a. To include the final and approved mechanical load fed by the bus duct for bus duct sizing.
    b. To ensure no sprinkler system should be installed over the bus duct.
17. Isometric drawing based on the approved structural and architectural drawings showing the complete bus duct with all accessories, complete route with direction changes, lengths of all parts, fixing method, location of tap off in each floor, cable connection to the tap off with the correct bend due to cable size and final connection to MDB or Transformer should be provided by the supplier and the contractor.
18. A flow chart for all bus duct components (Adaptors to the MDB and transformer, bus duct piece lengths, tap off and location, cable size) will be provided.
19. Fire barrier around the bus duct through walls to be submitted for approval, the used fire barriers to be submitted based on the manufacture recommendations.
20. The material submittal should be based on the approved load schedules, schematic, isomeric and single line diagrams by the consultant and the Authority for the complete bus duct.
21. The bus duct system should be tested and inspected in the factory with a consultant's presence.
22. Valid supplier trade license with expiry date, and ministry of trade certification as a legal supplier of the material in the submittal.
23. Authority approval for the bus duct on workshop drawings should be provided before ordering the material.

Notes:

1. **Accurate measurements are very important for smooth and fast installations; hence it is advisable to call the supplier specialist to take the required measurement by himself to be responsible for measurement accuracy.**
2. **It is advisable that the last piece of the bus duct be ordered at the end of the project (with an agreement with the supplier) to send him the final dimensions if the last piece's dimension is not typical.**

Workshop drawing for the Bus Bar Duct.

The consultant should instruct the contractor to prepare the workshop and coordination drawings for the bus duct system (to be attached with power drawings) to be submitted for the consultant's approval. The following should be considered in the workshop drawings:

1. Consultant and Authority approval for bus duct system shop drawings and isometric diagram, including the approved load schedules and voltage drop calculations.
2. Submission for approval for main and sub-main electrical rooms workshop and coordination drawings that include the following:
   a. Separate coordination drawing for all activities inside main and sub-electrical rooms including all details (plan and elevations) for all activities inside the room (switchgear equipment, routing of bus duct, sizes of openings, tap off and the cable to the SMDB entry, cable trays, A/C System, Switch Socket Outlet, Lighting Fixture etc....).
   b. Separate coordination plan drawing showing the bus bar duct rout with all services around it with sections (where necessary) with dimensions of all openings in slab/ walls according to manufacturer recommendations.
   c. Detailed drawing for all openings between floors/ areas showing the type of used fire resistance material and how to install it around the bus bar duct as per the manufacturer's recommendation.
   d. The supporting system shows the supporting details (between floors) and type of hangers going to be used (rugged or spring).
3. The prepared shop drawings should consider the authority's comments on the design drawings.
4. Final mechanical loads with consultant approval. (Chillers, pumps, fans, P.U., etc.).
5. Bus ducts shouldn't be routed through closed areas like stores or shops.
6. If the bus duct is changing the room by horizontal routing, the coordination drawing for that section to be provided showing all services, opening sizes, ensuring no contradiction with other services and easy future maintenance for all systems.

7. Shop drawings, isometric drawings to be signed by the supplier specialist of the bus duct. The supplier specialist signature confirms that the supplier specialist is accepting the installation details and it is matching the supplied system.
8. Workshop drawing should be submitted for Authority approval after being approved by the consultant.

Pre-installation requirements for bus duct

Points to be considered before starting installations:

1. When the procured bus duct reaches the site, the site engineer should check the received materials, ensuring that they match the approved submittal. The supplier should provide and submit relevant ASTA certificates and factory test reports for the bus bar and a checklist for inspection, ensuring the following as shown below:
    a. Rating
    b. Voltage
    c. No. of Bars
    d. Short Circuit Rating
    e. Rated to local Ambient Conditions
    f. Bus bar Configuration
    g. Type of Insulation
    h. Bus bar Type
    i. Bus bar Dimensions (Phase + Neutral + Earth)
    j. Earth Bar Bonding
    k. Bus bar to be enclosed in Insulating Material
    l. Overall, Bus bar riser Dimension
    m. I P Protection
    n. Fire Barrier
    o. Bus bar Supports.
2. The contractor should ensure full protection for the bus duct inside the stores.
3. The contractor should protect the bus duct when being installed and be protected from being hit due to other construction activities. The area near and around the bus duct should be kept clean from other services and to be protected from water. **See 6-9**

The 6-9 bus duct is covered with nylon for protection from site activities. The slab opening is covered by a suitable base plate for easy access to the tap-off unit and for safety in maintenance work.

4. Ensure full coordination with other systems connected to the bus duct (MDB, SMDB or transformer) considering the adaptor for connection, rating and actual load, routing of the bus duct, and location of the tap-off and tap-off ratings.
5. Supplier specialist approval for bus duct support system calculations and installation details should be provided according to the manufacturer's recommendations for all bus duct route, including vertical and horizontal routing.
6. Re-check the alignment of the wall and slab openings to ensure the correct location and size for straight vertical and horizontal bus duct alignment.

Installation requirements for construction

After sample approval for the bus duct (if the sample is submitted in the project), and based on the progress program of the project, the consultant should follow the contractor to start the preparations for the installation of the bus duct.

The notes shown below should be taken into consideration in the installation of the busduct:

1. The consultant should instruct the contractor to submit the Installation Method statement for the busduct for consultant approval. Installation work shouldn't start before the approval for the method statement or at least approval with notes that can be rectified within installation work.
2. Ensure the consultant approval for the workshop, coordination, and isometric drawings for the busduct system and ensure the authority approval for the shop drawings.
3. The bus duct system should be inspected in the factory with the consultant and the contractor engineer presence.

4. Supplier should provide the details and schedule for materials delivery after final approval, this schedule to be coordinated with site and civil/ MEP work to deliver the busduct in the correct time without delaying the work.
5. The bus duct should be installed and levelled correctly horizontally and vertically using the supporting system and then tightened to the correct torque.
6. The supplier and Contractor's Site Engineer should monitor the installation activity till completion. A periodic progress report showing site status, requirements, etc. Shall be submitted to the consultant. The supplier engineer should certify the installation, confirming it is healthy and fit for energization at the end of the installation. The following are to be ensured in the installations:
   a. The bus bar supporting channel on the wall should be installed according to the manufacturer's recommendations.
   b. The proper alignment for the bus bar during installation is to be considered.
   c. Switchgear/ Transformer supplier should supervise the installation/ connection of bus duct adaptor with switchgear/ Transformer and certify the connection as healthy and fit for energizing.
   d. All cables for the tap-off units are to be connected properly using a suitable crimping tool and installed, allowing flexibility in connecting and removing the tap-off. The tap-off cable is to be routed to the SMDB entry, which should match the site requirements.
   e. Adequate protection for bus bar ducts in their vicinity to the other works should be considered until the end of this work.
   f. Make sure that fire resistance material has been provided in wall openings after completing the installation.
7. Activity chart, work progress and look ahead program of work for the bus duct clarifying steps for executing the work (civil, electrical, finishing work, testing, and commissioning) and defining scope of work for the involved contractors with monthly updates.
8. Break down for all item related to the system shall be proposed by contractor for billing purpose; these broken-down percentages should be agreed by all the parties.
9. The contractor should provide the engraved labelling and identifications for the bus ducts, the tap and connected cables.
10. The earthing system for the bus duct is to be provided according to the manufacturer's and authority's recommendations.
11. Provision of the approved fire barrier around the bus duct after installation completion with a test certificate for the record.

Testing and commissioning of the bus duct system

When the contractor completes the installation of the busduct system and gets the approval from the consultant, the consultant should instruct the contractor to ensure and provide the followings before starting testing and commissioning:

1. The contractor should submit the T&C method statement for the bus duct for the consultant's approval.
2. As a pre-commissioning, the contractor should conduct a visual inspection and ensure the followings:
    a. The switchboard flange for the bus duct connection to the switchboard should be installed with the correct phase orientation, as shown in the manufacturer drawings and submittal.
    b. All bus duct elements are mounted according to the manufacturer's recommendations.
    c. All suspension brackets and spring hangers installed complied with the manufacturer's drawings.
    d. Suspension bracket clamps are not over-tightened and deformed.
    e. All joints cover plates are in position with rubber gaskets properly aligned preventing water or dust to penetrate inside the joint.
    f. All rubber bushes of joints cover plates are in position and properly aligned preventing water or dust to penetrate inside the joint.
    g. End covers are installed with the phase partitions between the phases' , neutral and earth bars.
    h. Vertical and horizontal run check (max deflection allowed is 2.5% of the total length).
    i. All joints' bolts are tightened properly and marked by suitable marker according to the manufacturer's recommendations.
    j. For plug-in tap-off boxes' installation on the feeder element, the Allen bolt on the top is tightened so that the clamps on either side are clamped on the body of the bus duct. The bottom screws are tightened to the correct torque.
    k. Tap off outgoing cable termination is smooth and tension-free.
3. Active commissioning: The following activities must be carried out only by qualified personnel and according to local electricity regulations where the bus duct riser is installed.
    a. Megger test of busduct riser (500V, 1000V): Disconnect the busduct and isolate the transformers, panel switches, panel measurement instruments, control fuses, shunts etc. that are connected in the switchboard. Ensure that all outgoing feeder MCCB, Switches and control fused are disconnected so that the voltage applied should not damage the connected devices.

       Perform the insulation resistance test for the complete line. If the resistance is more than 1000 Ohm/Volt for each circuit, then the result is OK; record all values in the test report.

b. Continuity test for the bus duct (500V, 1000V): Ensure that the bus duct is isolated completely, and check the continuity of each bar separately by an ohmmeter. The value should be nearly zero.

c. Earth continuity test: Ensure that the bus duct is isolated completely, and check the continuity of busbar earthing by an ohmmeter. The value should be nearly zero.

d. Load test: After completing the insulation test with successful results, all busduct components to be reconnected again for load test. All protective devices should be in operation for safety (ELCB and breakers). All outgoing circuits connected to the bus duct should be separated.

   Apply voltage to the busduct starting from the switchboard/ Transformer flange, then connect the outgoing circuits after applying the voltage one by one till full loading and check and ensure correct voltage and neutral is available in the end of the busduct.

e. Infrared testing with full loading: Verify the entire length of the busduct riser using infrared camera and capture the photos with full loading looking for heated spots. In case of overheated spots, disconnect the voltage immediately and remove the joint cover and check the joint or the feeder element for any non-conformity. If any non-conformity is found, then consult the supplier specialist for inspection and corrective action.

4. Final approval for the bus bar duct is subject to complete installation and successful testing and commissioning.

O &M manual for bus duct.

The consultant should follow the contractor to provide and submit the operation and maintenance manuals for the bus duct for consultant approval, the manuals should contain the following:

1. Description of the project system.
2. Clarifications about using the manual.
3. Troubleshooting guide for the system.
4. Safety instructions and requirements at work.
5. Start-up and shut down procedure.
6. An original catalogue highlighting the used material.
7. As built drawing including riser and Isometric diagrams
8. List of spare parts and special tools.
9. Procedure for routine maintenance for monthly/1 year/5 years...etc.
10. Authority NOC.

11. Warranty certificate for 2 years (as per project specifications).
12. Testing and commissioning certificates/reports.
13. Emergency information.
14. Completion certificate for the system from the supplier specialist
15. Manufacturer/supplier information and contact details.
16. Overview and method of operation (sequence of operation).
17. Training for operation and maintenance team. (It may require submission of training manual according to the project specifications).
18. All guidelines for using bus duct devices.

Switches and socket outlets

Requirements for Switches and socket outlets submittals.

The consultant should instruct the contractor to obtain the consultant's approval for the used materials, taking into consideration the following points:

- The manufacturer is listed in the consultant's approved manufacturer list.
- Compliance Statement is provided as per project specifications.
- The original catalogue is included in the submittal. (Highlighting all technical points, used components with part no, and requirements in the specifications).
- Submission of Samples or sample board.
- Valid supplier trade license with expiry date, and ministry of trade certification as a legal supplier of the material in the submittal.
- Method statement for installation.
- All used components should contain an Earth terminal.
- Separate submittal for all wiring devices related to all other low current systems ex: MATV, Intercom, telecommunication .. etc).
- A list for all wiring devices clarifying:
    a. Type (moulded or metal).
    b. Location.
    c. I.P. rating… etc.
- All double pole switches shall incorporate a pilot light and engraved identification like:
    a. Water heater (W.H.).
    b. Washing Machine. (W.M.).
    c. Dryer Machine (D.M.).
    d. Cooker (Cooker).
    e. Switch panel 3 phases (Danger 400V).
- Type of hook for the chandeliers (load calculations).
- All double Pole Switch for systems like intercom/Fire Alarm system etc. should be with pilot lamp and fuse.

- Documentary proof for (country of origin) for the submitted wiring accessories to be included in the submittal.

Switched and socket outlet requirements after material approval:

1. The contractor should ensure all material arrived at the site have passed MIR (material inspection request) and are approved by the Consultant.
2. The main contractor will ensure protection for materials inside the store and after installation at the site.
3. Contractor must ensure that the correct revision of the approved workshop/ coordination drawings is being used. He should note that any clouded areas on the drawings are subject to revision, hence work should not proceed in these areas until an updated drawing has been issued and approved.
4. The contractor must ensure that all safety requirements are fulfilled, including suitable accessibility to the site. The site should always be cleaned and dried in accordance with the project safety plan.
5. Contractor must use the approved method statement of installation and fixing the material to get the best quality of workmanship.
6. Breakdown for all items related to the materials shall be proposed by the Contractor for billing purposes, such breakdown should be agreed by all parties.
7. Activity chart and program of work for fixing switch and socket-outlet shall be submitted. The program should include steps for executing the following work:
    a. Civil work,
    b. Electrical work,
    c. Finishing work,
    d. Testing and Commissioning.
8. Engraved labels and identification should be provided as per the approved submittal of the material.

Points to be considered before fixing outlets.

a) Ensure that the edges around all switches and sockets are done properly.

b) Ensure that the paint around switches and sockets is completed, or at least the first layer is completed.

c) Ensure all J.B. (Junction boxes) and all end wires are clean.

d) All shaver sockets should be part of the power circuit (not the lighting circuit) as per Authority requirements to fit with a low-voltage transformer.

e) Neutral wires should not pass through the G.I. Box for switches and double pole switches.

f) Check that no bare copper should be visible on any terminal.

g) Check that no connection or tap is inside the G.I. box/D.B. box.

h) Check the availability of phase barrier in case of more than one phase is in the switch panel.

i) All outlets for the exhaust fan are to be kept 10-15 cm above the false ceiling. In case there is no space, the outlet should be on the ceiling or wall surface mounted in an accessible area for future maintenance. **See Photos 6-10**

6-10 Switches or supply points for equipment over the false ceiling should be close to the service opening for easy maintenance.

j) Ensure that earth wires are connected to the brass earth terminal of the G.I./box.

k) The surface installations, exposed external installations or surface installations in service areas for G.I. or PVC outlet boxes should consider the following:

   a. For PVC surface installed boxes, PVC conduit should be used. (PVC surface installation only over false ceiling based on authority regulations).

   b. For G.I. surface installed boxes, G.I. conduit should be used.

   c. For the W/P outlet, the W/P box should be used.

l) Ensure that all locations and heights of Junction and G.I. Box inside kitchens follow the coordination drawing and approved kitchen tiles sketch. Boxes in tiles in kitchens and bathrooms are to be in the middle of the tile.

Installation Procedure for Switch Socket Outlet

1. The wiring accessories will be mounted on the G.I. boxes fixed in the building fabric.
2. The accessories will be fixed using screws supplied by the manufacturer.
3. Complete the Wiring for the accessory in the approved way using the correct type of lugs and crimping tool.
4. Install identifications on the wires showing circuit No.
5. Accessories will be fixed in finished areas once the final paint is completed. (In case of delayed paint, the socket or switch can be fixed within its nylon cover with the first paint layer only, so the wall can be painted with the switch protected by its cover. **See photo 6-11**

6-11 Sockets and switches to be fixed with its protective nylon cover to keep it clean from dirt till project handover.

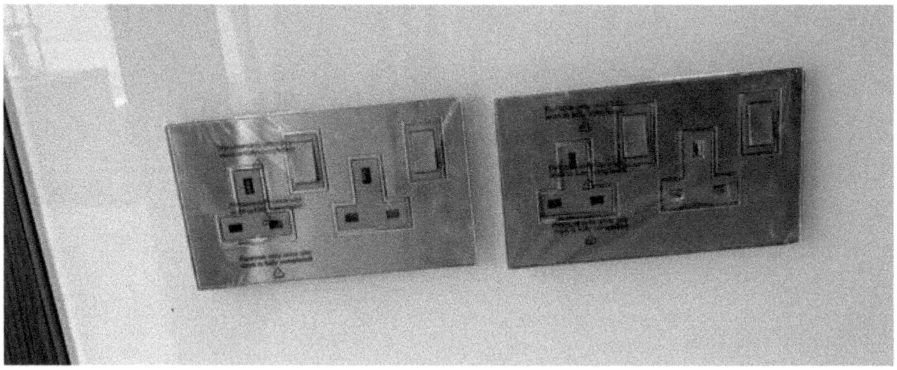

6. After fixing the outlet.
    a. Screws for all outlets must be convex in shape.
    b. The switch/ socket must be aligned in the horizontal direction.
    c. Ensure the location of all power outlets for the low current systems (CCTV, ACCS, EM system, etc.) as per the supplier's requirement and approved workshop drawing.
    d. Screw caps are fit to switches/sockets.

Operation and maintenance manual for Switches and Socket Outlets.

1. Description of the project system.
2. How to use the manual.
3. Original catalogue with highlights on the material used.
4. As built drawing including the wiring diagrams.
5. List of spare parts.

6. Procedure for routine maintenance for monthly/1 year/5 years.
7. Warranty certificate for 2 years (as per project specifications).
8. Testing and commissioning certificates/reports.
9. Manufacturer/supplier information and contact details.
10. All guidelines for using all devices/controllers, etc.

Chapter 7

Installations (Part 2) LV Distribution Systems

This Chapter is going to discuss the Low Voltage Distribution System, including:

1. MDB.
2. SMDB.
3. Final D.B.
4. Power Factor Correction System.

Approval of material submittal for LV Distribution System

Common requirements for MDB, SMDB, D.B. and Cap Bank.

The consultant should instruct the contractor to arrange periodical meetings between the Electrical engineer and the suppliers of LV Switchgear and all interlocked or interfaced systems to discuss and clarify the operation philosophy, modes of operation, material submittal requirements, Authority approval, work progress, actual mechanical load requirement, Load Schedules, and Schematic diagram, etc. The contractor should provide the following for the consultant's approval:

1. The supplier of the LV Distribution system has been listed in the consultant's Approved Manufacturers List.
2. Description of the LV distribution system and capacitor bank with modes and philosophy of operation.
3. Compliance statement to be submitted. The supplier should clearly show his compliance with the specifications, and non-compliant sections should also be highlighted with justifications.
4. The submittal should include the type test certificates for the panel sizes in the project.
5. The submittal should include the original catalogue for each component in the distribution boards, highlighting the technical points that exist in project specifications and highlighting the part number of the proposed components.
6. Valid supplier trade license with expiry date and Ministry of Trade certification as a legal supplier of the material in the submittal.
7. The material submittal should be based on the approved shop drawings, load schedules, schematics, and single-line diagrams by the consultant and the Authority.
8. Cable entry and outgoing to and from all switchgear equipment (MDB, SMDB, D.B. and C.B.) should be decided, whether from top or bottom, according to the approved workshop and coordination drawings and site conditions.
9. Ingress protection (I.P.) rating for the distribution panels should be suitable for the location of the board installation and should be according to the approved coordination drawings.
10. All indication lamps on the panel should be L.E.D. type or low-voltage lamps with built-in transformers.
11. Short circuit ratings for the breakers in all D.B.s should be according to the approved drawings, approved design calculations, or the authority regulations and recommendations.
12. All busbars inside the panels should be solid copper bars (multistrip laminated are not accepted); environmental conditions for the project and the panel should be considered in bus bar sizing.

13. Door hinges should be robust metallic hinges. All doors should be earthed and connected to the earth bar. Doors should be open at least 90 degrees.
14. If the capacitor bank is attached and part of the MDB, then a G.I. partition should be provided between them.
15. Flow chart for all switchgear equipment (MDB, SMDB, D.B. and C.B.) clarifying incoming and outgoing breakers, sizes of breakers, cable sizes, cable entry and cable access, location, length of cables, etc...., including any interface or connection to generator, B.M.S., F.A.S., P.M.S. etc....
16. Activity chart, work progress and look ahead program of work for LV switchgear clarifying steps for executing the work (civil, electrical, finishing work, testing, and commissioning) and defining the scope of work for the involved contractors with monthly updates.
17. Engraving labels and ferrules should be used to identify distribution system cables and wires, also, for inside components and outside on the boards.
18. Bus bar identification with R.Y.B. stickers and colours.
19. Except for final D.B.s, suitable internal lighting and an anticondensation heater with a thermostat should be provided.
20. The final and approved mechanical loading (mechanical equipment ratings, pumps, compressors, fans, etc..) by the consultant. This will enable correct breaker and bus bar sizing.
21. All distribution and power factor correction panels should have doors covering the internal circuit breakers. Exposed breakers are not allowed.
22. I.P. ratings for the distribution system panels should be according to the specifications and location of installation.
23. The distribution system should be tested and inspected in the factory in the presence of a consultant.

Specific requirements for Main Distribution board material submittal.

1. Ratings and sizing of the protection devices (breakers) should be according to the approved drawings and the final approved mechanical loads.
2. All MCCBs shall be the electronic type with adjustable settings for overload and short circuits and with the facility to test the tripping by using an electrical pulse from a portable unit.
3. All busbars in MDBs should be solid copper wire, (laminated multistrip bus bars are not permitted) sized with 20c degree above specified temp. (sized for 70c).
4. The form of construction for the MDB should be according to the specifications and approved drawings.

5. Air circuit breakers should have the following protections, short circuit, overload, undervoltage, earth fault.
6. The circuit breaker feeding the D.B. inside the transformer room from MDB should be of the current limiter type.
7. Measurement instruments should be according to the authority requirements and should include at least a K.W.H. meter, Ammeter, voltmeter, power factor meter, etc.
8. Test links should be provided in the neutral bus bar.
9. Chiller MCCB's control circuit should include the following protections: phase failure relay, reverse phase relay, under voltage relay and over current relay.
10. Surge arrestors should be provided for lightning protection.
11. Detailed calculations for busbar support and spacing should be provided for approval.
12. MDB internal wiring should be fire-rated type wires of 105c with R, Y, B and Y.G. colours.
13. Anti-condensation heater and humidity sensor should be provided.
14. Samples should be provided if applicable.

Specific requirements for Sub-Main Distribution board material submittal.

1. ELCB and MCCB ratings and sizing should be according to the approved workshop drawings and load schedules.
2. Door openings for the SMDBs should be decided to be left, right, or 2-leaf doors based on the coordination drawing and the available space in the room.
3. All MCCBs should have thermal and magnetic protection and tripping units according to the approved drawings and specifications.
4. The main incomer MCCB should be of lockable handle type.
5. All busbars in SMDBs should be solid copper wire, (laminated multistrip bus bars are not permitted) sized with 20c degree above specified temp. (sized for 70c).
6. All external SMDBs should be provided with a suitable outdoor canopy.
7. SMDBs internal wiring should be fire rated type wires of 105c with R, Y, B and Y.G. colors.
8. Samples are to be provided, if possible, for final approval.

Specific requirements for final Distribution board material submittal.

1. For service, D.B. components and requirements, consider the lighting system mode of operation in

service and wide areas as it may require additional controls like a timer, contactor, etc.

2. Color of distribution board enclosure in front of house should be according to the project specifications, Architect, and client approvals.
3. The short circuit level for all M.C.B.s should be 10KA minimum or according to the approved calculations or the authority recommendations.
4. Undervoltage relay protection should be used for all breakers feeding A.C. loads.
5. Stoppers should be provided at both ends of the rows; blank covers should be provided for all spaces between breakers.
6. D.B. hinges should be metallic type.
7. Similar Protection level for the breakers and their related contactors.
8. Cookers and heavy loads (split A.C., water heaters and window A.C.s) should be distributed and rotated between the phases in the flats, floors, and houses to achieve load balancing.
9. For D.B.s with more than one section, the neutral bar should be separate for each section.
10. At a minimum, there should be 2 Earth bars in the D.B.
11. ELCB and its related breaker for an individual section should be in one row.
12. Ratings for all breakers should be according to the final approved load schedules, including the actual mechanical loads and other heavy loads.
13. Ensure recess enclosure for recess installed Distribution boards. Back boxes for the D.B. should be provided in early stage to be inserted in the block work and connected to the conduits.
14. Miniature circuit breaker tripping characteristics should be selected, whether type B, C or D, according to the load type (power, lighting, or motor).
15. The number of D.B. sections and amperage rating for the circuit breaker and earth leakage circuit breaker should be determined according to the consultant and authority's approval.
16. If the incomer cable is armoured and connected to a recessed D.B., then a gland box is to be provided at the top or bottom of the D.B. based on cable entry to allow glanding and armour earthing.
17. D.B.s internal wiring should be fire rated type wires of 105c with R, Y, B and Y.G. colors.
18. Under voltage relay should be provided for all A.C. equipment (check the specification and authority regulations).
19. Samples should be provided for final approval.

Specific requirements for Capacitor Bank material submittal

1. Capacitor bank sizing calculations should be submitted for consultant approval; sizing of capacitor banks should be according to the consultant and authority approval for workshop drawings and load schedules.
2. The submittal should include the original catalogue for each component in the distribution boards, highlighting the technical points that exist in project specifications and highlighting the part number of the proposed components.
3. A clear and clarified sequence of operations is to be submitted.
4. The capacitor banks should include fast discharge resistors. Contactors should be sized to withstand the inrush currents.
5. Ensure good ventilation inside the capacitor bank to avoid heat generation.
6. The type of fuse should be shown (AM, G.H., GDGH), and the fuse handle should be provided for fuse replacement.
7. A number of switching steps should be determined according to the actual type of load.
8. Shunt coils for harmonic filters should be provided, and calculations should be submitted for consultants' approval.
9. A reactive power regulator with a capacitor step indicator lamp shall be mounted on the door of the cubicle.
10. IF the capacitor bank is part of the MDB, then the G.I. sheet is to be provided as the separation between them.
11. Samples are to be provided, if possible, for final approval.

Shop Drawings for Distribution System and capacitor banks

The consultant should instruct the contractor to prepare the workshop and coordination drawings for the Distribution and power factor correction systems to be submitted for the consultant's approval. The following should be considered in the workshop drawings:

1. Consultant and Authority approval is required for the distribution and correction power factor systems shop drawings, including the load schedule and mechanical loads.
2. Consultant and Authority approval for the coordination drawings for the main electrical room.
3. Approval of all mechanical loads. (Chillers, pumps, fans, P.U., etc.).
4. Ensure load balancing between phases according to the Authority regulations and requirements.

5. Coordination drawings (plan and elevation) in details for all Main and Sub Electrical rooms should be submitted for consultant and authority approval considering all activities inside the room (switchgear equipment, routing of cable tray, A/C System, Switch Socket Outlet, Lighting Fixture etc....). Taking into consideration size of door of Main and Sub Electrical room with reference to the distribution panels' size.

6. Heavy loads like electric cookers and window/ split A.C.s should be distributed on the phases between flats on the floor, and between houses in the compound to ensure balanced loading on all phases, D.B.s to be classified based on heavy load distribution so in the installation of the D.B. core, selection of the D.B. core for the specified flat of house can be easy depending on the phases of the heavy loads.

Installations of Distribution System

After sample approval for (DBs, SMDBs, MDB and C.B.), and based on the progress program of the project, with Containment work completed, the consultant should follow the contractor to start the preparations for the installation of switchgear equipment.

General notes for Installation of the distribution system

Below are general notes to be considered in the installation of the distribution system panels:

1. The consultant should instruct the contractor to submit the installation Method statement for the distribution system for the consultant's approval. Installation work shouldn't start before the approval for the method statement or at least approval with notes that can be rectified within installation work.

2. All mechanical loads are finalized and submitted officially to the electrical engineer to finalize the D.B.s sizing. Also check the rest loads on the panels and ensure the supply branch and the feeding cable is suitable for the load.

3. Ensure the authority approval of the shop drawings.

4. Ensure the consultant's approval for the workshop and coordination drawings for distribution and power factor correction systems.

5. Distribution panels and capacitor banks should be inspected in the factory with the presence of the consultant and the contractor engineers.

6. The supplier should provide the details and schedule for materials delivery after final approval; this schedule should be coordinated with the site, civil and M.E.P. work to deliver the panels at the correct time without delaying the work.

7. Proper protection for all devices, wires, components, panels ...etc. shall be provided during storing and installation.

8. All keys for the panels and the doors to be kept in a suitable key box with proper and permanent identification on each key.
9. Identifications, markers, engraved labels, ferrules, and warning sign stickers should be fixed at appropriate locations in distribution panels.
10. All cuts in the plate for cable insertion should be painted with zinc paint to prevent future rust.
11. The panels should be installed and correctly levelled after tightening all bolts with washers and nuts.
12. Supplier, Contractor's Site Engineer should daily and continuously monitor the installation activity till completion. Periodical progress report showing site status, requirements ...etc, shall be submitted by the contractor.
13. All earthing wires in the panel should be connected to the earthing bus bar in the panel. Earthing busbar should be connected to the earthing system (earthing rod).
14. All wirings and cables should be connected properly using a suitable crimping tool.
15. Break down for all item related to the system shall be proposed by contractor for billing purpose; these broken-down percentages should be agreed by all parties.

Specific notes for MDB Installations

The following activities need to be confirmed and checked by the contractor before and during the MDB installation work:

1. Consultant approval for coordination drawing (showing all required dimensions) for the main electrical room that MDB will be installed in.
2. MDB accessories and components (MCCB, Isolators, ELCB, contactor, timer, bus bar, etc.) should be kept clean from dirt and dust and free from mechanical damage. Ensure no tools are forgotten in or on the MDB after work completion.
3. Water pipes over or behind the MDB in the Electrical room are totally prohibited.
4. The I.P. rating and form of construction should be according to the approved material submittal.
5. All door hinges should be metallic.
6. MDB installation should allow easy access to the control and cable sections, whether the cable section is front or rear access. Doors should be able to open at least 90 degrees. Front access should allow withdrawable A.C.B.s to be easily removed from the panel. Spaces and dimensions around MDB are according to the authority requirements that allow easy movement, easy maintenance and component replacement.
7. Ensure proper painting for the MDB that protects the panel from rust.

8. Cables and wires to be tested for insulation by suitable megger before glanding and wiring addressing. Glands should be earthed and connected to main earthing terminal.
9. All cables routed through G.I. conduits shall be inserted into the MDB through a G.I. adapter.
10. The bus bar shall be secured/ tightened to the correct torque using a torque spanner. They should be properly shrouded. All screws should be tightened to the correct torque using a torque spanner, screws should be marked after tightening.
11. Ensure correct connections to the measurement Instrumentations (ammeter, voltmeter, etc.) and properly tightened.
12. For the K.W.H. meter branch, a padlocked circuit breaker (MCCB) should be used to control the power to the meter as recommended by the authority.
13. MDB panel earthing should be connected to the earth bar inside the MDB.
14. MDB width should be more than the corresponding trench. MDB fixing on ground should follow the MDB submittal and catalogue.
15. If the main electrical room is on a floor different from the main low voltage panel, then an emergency isolation facility (emergency push button) is required to be installed in a suitable position to be reached and noticed on the ground floor.

Specific notes for SMDB Installations

The following activities need to be checked by the contractor during the SMDB installation work:

1. Consultant Approval for coordination drawing (showing all required dimensions) for the submain electrical room that contains the SMDB. The final location for the SMDB installation will be decided accordingly.
2. SMDB accessories and components (MCCB, Isolators, ELCB, contactor, timer, bus bar, etc.) should be kept clean from dirt and dust and free from mechanical damage. Ensure no tools are forgotten in or on the SMDB after work completion.
3. Water pipes over or near the SMDB in the Electrical room are totally prohibited.
4. The SMDB location should allow easy access. SMDB doors should be opened at least 90 degrees for maintenance.
5. Before glanding and wiring addressing, an insulation test for the cables and wires should be completed. Glands should be earthed and connected to the main earthing terminal.
6. The SMDB should be installed according to the approved installation drawings. The installation will be by suitable-sized channels to be fixed according to the approved height and location, fixed by threaded

rods fastened with suitable washers and nuts.

7. All cables routed through P.V.C. conduits should be inserted into the SMDB through the P.V.C. adapter.
8. The bus bar shall be secured/ tightened to the correct torque using a torque spanner and properly shrouded. All screws should be tightened to the correct torque using a torque spanner, and screws should be marked after tightening.
9. Ensure correct connections to the measurement Instrumentations (ammeter, voltmeter, etc.) and tightened properly.
10. Cables and wires should be connected to the circuit breakers through suitable lugs and should be tightened correctly to the specified torque; nuts should be marked after tightening.
11. Final installation approval is subject to the successful completion of the testing and commissioning of the SMDB and authority approval.
12. The panel should be installed correctly by suitable fixing bolts, ensuring straight horizontal levelling.

Specific notes for D.B. Installations

The following activities need to be checked by the contractor during and after the D.B. installation work:

1. Detailed coordination sketch for D.B. location with sections and elevation should be submitted for consultant's approval in advance before construction of block wall. These coordination sketches should show and clarify all dimensions, levels, and installation details.
2. Check the wall for the D.B. installation, the thickness should be at least 20 Cm to fix the D.B. in, if the wall is less than 20 Cm thickness, coordinate with civil work to change the thickness to 20 Cm.
3. D.B. I.P. rating should be according to the approved submittal based on the location of installation; D.B. colour should match the approved colour for the sample D.B. by the client and the Architect.
4. Final D.B.s' accessories and components (MCCB, M.C.B., Isolators, ELCB, etc.) should be kept clean from dirt and dust and free from mechanical damage. 2mm plywood shall be used to cover the enclosure immediately after installation to avoid cement from spoiling the D.B. enclosure.
5. D.B. should not be installed where a water pipe is passing above or behind.
6. If an F.C.U. Above, the false ceiling is located near the D.B., the level of the drip tray should be positioned away from the D.B. location to prevent water from dripping on the D.B. **See photo 7-1.**
7. For recessed D.B.s under the false ceiling, Add extra recessed P.V.C. conduits (at least 2) from the D.B. to over the false ceiling for future requirements.
8. The wall holding the D.B. should be filled with concrete in the D.B. location and around the D.B. to allow secure fixing and installation.

7-1 Drip trays under the fan coils to prevent dripping on the electrical panels and equipment.

9. All cuts in the plate for cables or wire insertion should be painted with zinc paint to prevent future rust. For cables use gland to protect the cable from the sharp edge, for wires, use rubber or P.V.C. grummet to protect the wires near the cut edge.
10. The D.B. should be installed in an accessible area. D.B. shouldn't be installed in private areas or outside.
11. Concrete around the back box and D.B. Edges should be made well during plaster work to eliminate any holes or cavities around or behind the D.B.
12. All P.V.C. conduits connected to D.B. should be through a P.V.C. adapter.
13. Insulation test for all wires to be done before dressing.
14. Proper crimping for wires of 6mm size and above.
15. Ensure that all electrical connections are correct and tightened properly to the corresponding devices.
16. Shroud (P.V.C. insulation) shall be done on all five wiring terminals (Red, yellow, blue, black and yellow-green).
17. If the house or the building contains more than one floor, then each floor should have its own D.B., according to the authority recommendations, the D.B. in one floor shouldn't feed the second floor.
18. The D.B. should be installed to allow the door to open more than 90 degrees.
19. For armoured cables, an earth connection between the gland and the main earth terminal shall be provided.
20. Single P.V.C. cables for the 3 phases, neutral and the earth feeder entry connections can be used if cables are routed through P.V.C. conduits from the SMDB to the final D.B. No gland is required. If

armoured or XLPE cables are used laid under the gourd from the SMDB to the D.B., then for recess D.B.s, an additional section in the D.B. acting as a Gland Box is required before the entry section of the cable to allow glanding of the cable on the D.B. plate. The Gland box should be covered same as the D.B.

21. The D.B. should be installed correctly by suitable fixing bolts, ensuring straight horizontal levelling.
22. When installing the D.B. breakers and bus bars after installing the enclosure in the wall, ensure the correct type of D.B. core is used based on the phases of the heavy loads (cooker, A.C., and water heater) as these loads should be altered sequentially in the floor flats or in the houses in residential projects.
23. As built, the load schedule is to be fixed on the D.B. door inside the D.B.
24. Washer for all screws to be provided on the internal cover of the enclosure.
25. Final installation approval is subject to the successful completion of the testing and commissioning of the D.B. and authority approval.

Specific notes for Capacitor Bank Installations

The following activities need to be checked by the contractor during the Power Factor Correction System installation work:

1. Approved coordination drawing (showing all required dimensions) for the electrical room containing the Capacitor banks. The drawing should be approved by the consultant, and then the final location for the capacitor bank installation will be decided accordingly.
2. Capacitor bank accessories and components (MCCB, Isolators, H.R.C. fuses, contactors, timer, bus bar, etc.) should be kept clean from dirt, dust, and mechanical damage. Ensure no tools are being forgotten in or on the cap. Bank panel after work completion.
3. Water pipes over or near the capacitor banks in the Electrical room are totally prohibited.
4. The capacitor bank location should allow easy access. Doors should be opened at least 90 degrees for maintenance.
5. Cables and wiring insulation should be tested before glanding and wiring addressing. Glands should be earthed and connected to the main earthing terminal.
6. All cables routed through PVC/GI conduits should be installed through PVC/GI adapter.
7. The bus bar shall be secured/ tightened to the correct torque using a torque spanner and properly shrouded. All screws should be tightened to the correct torque using a torque spanner, and screws should be marked after tightening.
8. Ensure correct connections to the measurement Instrumentations (ammeter, voltmeter, etc.) and properly tightened.

9. The panel should be installed correctly with suitable fixing bolts, ensuring horizontal levelling based on the supplier's recommendations for fixing.
10. For ground mounting installations, the width of capacitor banks and power factor correction panels should be more than the corresponding trench. Panel fixing on the ground should follow the Power factor correction panel installation details in the submittal and catalogue. Clearance spaces around the panel should allow easy movement and replacement of the biggest equipment in the room or should be according to the authority regulations.
11. Final installation approval is subject to the successful completion of testing and commissioning of the power factor correction system and authority approval.

Testing and commissioning of Distribution System

General Notes for T&C for distribution system.

When the contractor completes the installation of the distribution system and gets the approval for system installations for (DBs, SMDBs, MDB and C.B.). from the consultant, the consultant should instruct the contractor to ensure and provide the following before starting testing and commissioning of switchgear equipment.

1. The T&C method statement should be submitted to the consultant for approval.
2. Switchgear installations and protection are completed and complied with the authority requirements as shown below:
    a. Protection against direct contact by termination for all earthing and bonding conductors to the earthing system.
    b. Easy access to all switchgear equipment.
    c. Identifications and danger notes/ warnings are fixed on the doors.
    d. All wiring connections are connected correctly and tightened.
    e. Cable routing and connections are completed and mechanically protected. Selected cables and wires having correct current carrying capacity and voltage drop according to the approved workshop drawings.
    f. Wiring accessories (Switches, sockets, lights, bell, cooker etc.) connections are done and completed correctly.
    g. Fire barrier materials are fixed.
    h. As built drawings, load schedules and single line diagrams are provided inside the switchgear and panels.
    i. Outdoor canopies are provided for external panels.
3. Submission of Testing & Commissioning program by the Contractor, Supplier or Third party as applicable for consultant's approval, including the following:
    a. Letter-headed testing template with the agreed format for all tests to be provided according to the specifications and applicable standards showing successful criteria for the test results.
    b. List of testing instruments to be used with valid calibration certificates.

Testing and commissioning of final D.B.

Before starting the DB T&C, the consultant should instruct the contractor to complete all work related to the final D.B.s installations and circuit connection, like wiring, fixing the wiring accessories, and connection of mechanical equipment (A/C, W/H, Ex fans, etc.). The followings should be checked in the DB T&C activities:

1. The contractor should submit the T&C method statement for the consultant's approval.
2. The contractor should conduct a visual inspection and ensure the followings:
    a. For all electrical components, such as fittings, socket outlets, etc., that are connected to the final D.B., ensure the fittings are installed properly without any damage due to construction activities.
    b. Check the internal distribution board for C.B. ratings, internal wiring, feeder cable, electrical and mechanical connections, identifications, labelling etc., are all correct and match the approved load schedules.
    c. All metallic components are properly earthed and bonded.
3. Polarity check: ensure all electrical components and fittings are connected correctly to the corresponding terminals on the fitting.
4. Function test: Complete the following:
    a. Megger test for the D.B. (disconnect sensitive devices).
    b. Earth leakage test (ELCB).
    c. Mechanically operate the circuit breakers to switch them on, off, and trip.
    d. Megger test for the incomer and outgoing cables.
    e. Plug in the test for all 13A and 15A switched socket outlets.
    f. Operation of all circuits (lighting, power, A.C., W/H, etc.) connected to the D.B. (may require a power supply or generator).
5. Final approval is subject to the successful completion of the DB T&C activity.

Testing and commissioning of SMDB.

Before starting the SMDB T&C, the consultant should instruct the contractor to complete all work related to the final D.B. installations and SMDB outgoing circuits. The following to be checked in the SMDB T&C activities:

1. The contractor should submit the T&C method statement for the consultant's approval.
2. The contractor should conduct a visual inspection and ensure the following:
    a. For all electrical outgoings from the SMDB to the D.B.s, control panels, disconnecting switches, etc., with their cables that are connected to the SMDB, ensuring the cable and the load are installed properly without any damages due to construction activities.
    b. Check the MCCB ratings of the SMDB, internal wiring, feeder and outgoing cables sizing, electrical and mechanical connections, identifications, labeling etc. are all correct based on the approved shop and coordination drawings (load schedules and wiring diagram).
    c. All electrical connections are correct and tightened.
    d. All electrical connections are shrouded and not damaged.
    e. All fuses and links are inserted in their places.
    f. All C.T.s are rated correctly and properly connected.
    g. All control interfaces are installed correctly.
    h. All control wiring are installed correctly, labelled clearly, and segregated from the main power wiring.
    i. All equipment and components are labelled correctly.
    j. All fuse ratings, Protection device settings, indicator lamps, conductors, voltmeter, ammeter, and selector switches are as per the approved drawing (load schedule and wiring diagram) and submittal. Check for proper operation.
    k. Phase sequence and bus bar/ cables phase colouring are correct and have identifications.
    l. Wiring and schematic diagrams are provided inside the panels.
    m. Earthing conductors and glands are connected to the earth bar.
3. Do the function test, including the following:
    a. Megger test for SMDB, incomer, and outgoing cables.
    b. Earth leakage test for ELCBs.
    c. Operation test for all MCCBs and contactors (Mechanical on /off/ trip)
4. Final approval is subject to the successful completion of the SMDB T&C activity.

Testing and commissioning of MDB

Before starting the MDB T&C, the consultant should instruct the contractor to complete all work related to the SMDB installations and all MDB outgoing circuits. The followings should be checked in the MDB T&C activities:

1. The contractor should submit the MDB T&C method statement for the consultant's approval.
2. The contractor should conduct a visual inspection and ensure the followings:
    a. All electrical outgoing connections from the MDB to the SMDBs and any other loads, control panels, disconnecting switches, KWhMETER D.B.S., etc., are completed. All incoming and outgoing cables to the MDB are installed properly without any damage due to construction activities.
    b. Ratings of all circuit breakers ACB/MCCB/ELCB/contactor are correct as per the approved schematic diagram/ load schedule and riser diagram.
    c. All electrical connections are correct and tightened.
    d. All electrical connections are shrouded and not damaged.
    e. All fuses and links are inserted in their places.
    f. All C.T.s are rated correctly and properly connected.
    g. All control interfaces are installed correctly.
    h. All control wiring are installed correctly, labelled clearly, and segregated from the main power wiring.
    i. All equipment and components are labelled correctly.
    j. Phase sequence and bus bar/ cables phase colouring are correct with identifications.
    k. Wiring and schematic diagrams are provided inside the panels.
    l. Earthing conductors and glands are connected to the earth bar.
    m. All fuse ratings, Protection device settings, indicator lamps, conductors, voltmeter, ammeter, and selector switches are selected according to the approved drawing (load schedule and wiring diagram) and approved submittal. All mentioned components should be checked for proper operation.
    n. Anti-condensation heater is functioning properly.
3. Do the function test, including the following:
    a. Megger test for MDB, incomer, and outgoing cables.
    b. Earth leakage test for ELCBs.

c. Ensure proper operation for all Circuit breaker contactors (Mechanical on /off/ trip)

d. Check that castle locks (if available) are installed, identified correctly, and operating properly.

e. Function test and operation of indication devices and selector switches, and check the continuity of power circuits.

4. Final approval is subject to the successful completion of the MDB T&C activity.

Testing and commissioning of Capacitor Bank and power factor correction panel

Before starting the Cap, Bank T&C, the consultant should instruct the contractor to complete all work related to the Cap. Bank and power factor correction system installations. The following is to be completed and checked before starting the cap. Bank T&C activities:

1. The contractor should complete the T&C for the MDB/ SMDB and the final D.B.s before the cap bank T&C...

2. The contractor should submit the Power factor correction system and Cap, Bank T&C method statement for the consultant's approval.

3. The contractor should conduct a visual inspection and ensure the following:

   a. The panel and the incomer cable from MDB are installed properly and without any damage due to construction activities.

   b. Check ratings of isolator, H.R.C. fuses, contactors (capacitor switching), reactors, capacitors are correct and matching the approved single line diagram.

   c. Check the current transformers, exhaust fans, thermostat, and power factor control relay inside the panel have been installed and wired correctly according to the approved drawings.

   d. All electrical connections are correct and tightened properly to the correct torque and marked.

   e. All terminals are shrouded properly without any cuts or damages.

   f. All control wiring are installed correctly, labelled clearly, and segregated from the main power wiring.

   g. All equipment and components are labelled correctly. Cable sizing is selected correctly.

   h. Phase sequence and bus bar/ cables phase colouring are correct and have identifications.

   i Wiring and schematic diagrams are provided inside the panels.

   j. Earthing conductors and glands are connected to the earth bar.

4. Do the function test, including the following:

   a. Megger test for bus bars before energization and insulation test according to the normal standards.

b. Check (On/ off) Mechanical operation of the isolator.

c. Check power factor switching in manual mode and ensure proper operation.

d. Ensure the Anti-condensation heater is working properly.

5. Final approval is subject to the successful completion of the power factor correction T&C activity.

O & M Manuals for Distribution System (MDB, SMDB, DB, Cap Bank).

1. Description of the project system.
2. Clarifications about using the manual.
3. Description of the interfaced systems with Power Distribution Systems like Generators, Fire Alarms and B.M.S. systems.
4. Troubleshooting guide for the system.
5. Safety instructions and requirements at work.
6. Start-up and shutdown procedure.
7. An original catalogue highlighting the used material.
8. As-built drawing including schematic, riser diagrams, interfacings, and load schedules.
9. List of spare parts.
10. Procedure for routine maintenance for monthly/1 year/5 years...etc.
11. Authority N.O.C.
12. Warranty certificate for 2 years (as per project specifications).
13. Testing and commissioning certificates/reports.
14. Emergency information.
15. Manufacturer/supplier information and contact details.
16. Overview and method of operation (sequence of operation).
17. Training for security, operation and maintenance team. (It may require the submission of a training manual according to the project specifications).
18. System programming, configurations, settings with the used software details and communication protocol.
19. All guidelines for using devices and controllers inside the control room.
20. Discrimination study and all protection breakers' settings.

Chapter 8

Installations (Part 3) Electrical Systems

This chapter discusses the following systems:

1. Lifts, Escalators and Travelators
2. Generators and A.T.S.
3. Earthing and Lightning Protection
4. Lighting Fixture and Lighting Control System

Lifts, Escalators, and Travelators

Material submittal approval requirements for lifts, escalators, and travelators.

The consultant should instruct the contractor to arrange periodical meetings between the Contractor's Electrical engineer, the supplier of Lifts, Escalators and Travelators and the suppliers of low current systems interfaced with the Lift, Escalators and Travelators like B.M.S., F.A.S., Intercom system, Access Control System, Emergency lighting, Public Address System, Switchgear and Generator, to discuss and clarify the installation requirements, connection to switchgear, load rating, material submittal requirements, modes of operation including operation in case of fire alarm or power supply failure, work progress and the latest authority (Civil Defense) requirements etc. The contractor should provide the followings for the consultant's approval:

1. The system supplier should be listed on the consultant's approved manufacturers list.
2. Description of the system (Lift, Escalator, and travelator) and all related components.
3. Compliance statement to be submitted. The supplier should clearly show his compliance to the specifications, non-complied sections to be also highlighted with justifications.
4. The submittal should include the type test certificates for the systems.
5. The submittal should include the original catalog for each component of the system, highlighting the technical points that exist in project specifications and highlighting the part number of the proposed components.
6. The submittal should include the Approved and signed workshop and coordination drawings for the lift shafts, Escalator and Travelators' plans and sections, showing clearly the finishing levels of floors and sections details with complete and clear dimensions, these drawings will be the base for supplying the systems and the supplier should sign the mentioned drawings confirming all requirements and details are Ok. Otherwise, he should highlight the required changes and return them to the contractor for modifications to adapt the mentioned systems properly.
7. The submittal should clarify the interfacing with the project systems (based on the project specifications) like B.M.S., F.A.S., Intercom system, Access Control System, Emergency lighting, Public Address System, Switchgear and Generator to clarify the followings:
    a. Interfacing with B.M.S.: The system should contain the following volt-free contacts in the control panel:
        i. Run status (on/off)
        ii. Alarm signal (healthy status)
        iii. Indication of Failure (faulty system).

iv. Lift position indicator

b. Interfacing with Fire Alarm System:

The system should be integrated with Fire Alarm System and should acting according to the Authority regulations in case of fire.

c. Interfacing with the Intercom system:

Lift supplier should provide the requirements for communication with outside intercom system, handset at control room/ reception desk.

d. Interfacing with Public Address System:

Lift suppliers should provide the requirements for communication with outside Public Address Systems.

e. Interfacing with Generator and Automatic Transfer Switch (A.T.S.).

Supplier of Lifts, Escalators and Travelators should provide the requirements for the approved mode of operation in case of power failure and power change over to A.T.S.

8. Block diagrams for all equipment/interphase with Lifts, Escalators, and Travelators are required (B.M.S., Fire Alarm System, CCTV, Power supply/schematic diagram for ESMDB feeding the systems, standby generator, isolators, Intercom, traffic director panel, A.T.S., alarm bell etc.
9. Details of earthing for the systems should be included in the submittal.
10. Identification with engraved labels and ferrules for all equipment cables and control wires should be part of the submittal.
11. An emergency landing device and safety shield should be part of the system.
12. The sequence of operation during the failure of the power supply/generator indicates the effect on the interfaced systems.
13. Evacuation procedure during power supply failure or any other fault or lift stop between two floors should be clarified in the submittal.
14. A voice annunciator in the lift car should be provided.
15. 24X24 System is required to be provided according to civil defence regulations.
16. The battery of the system should be maintenance-free.
17. The maintenance program for the system should be submitted.

Workshop drawing for Lifts, Escalators, and Travelators.

The consultant should instruct the contractor in cooperation with the system supplier to prepare the workshop and coordination drawings for the Lifts, Escalators, and Travelators systems to be submitted for the consultant's approval. The followings should be considered and prepared in the drawings:

1. Preparation of coordination and workshop drawings for lift machine room indicating the location of all equipment and other electrical/ mechanical and interfaced low current systems components, etc.. based on the project specifications and as shown below:
    a. The A/C system (F.C.U. unit and thermostat) should be installed away from the control panel or D.B.
    b. Control panel with system installation details.
    c. The main power supply is fed from SMDB, and a separate disconnecting switch should be provided for each lift.
    d. Lighting fixture, separate lighting switch, emergency lighting, switch socket outlet and F.A.S., etc...
    e. Alarm bell.
    f. All connections between low current systems and lift.
    g. All required trunking (should be on a high level), not floor level.
    h. Installation details and Location of board related to evacuation procedure that will be used in case of power supply failure or any fault when the lift stops between 2 floors.
2. Workshop drawing with full details and dimensions for shaft, including the followings:
    a. All lighting points inside the shaft with fitting type and two-way switch for operation.
    b. Alarm bell
    c. 13A S.S.O. at height 1.1m from pit level near the entrance side (for maintenance).
    d. All connections between low current systems and the lift indicate the route of all cables, which will run from the lift shaft to B.M.S., reception desk and control room).
    e. The finishing floor level should be mentioned clearly on the drawings for lift, escalator and travelator entrances..
3. Builders work for the lift shaft sizes and lift entrances showing the required openings in the walls and concrete for lift hall bush buttons and lift indication screen, the opening to be provided in coordination with civil work and as required by the system supplier.
4. All sleeves and conduits should be shown in the drawing for the system and the interfaced systems based on the systems' suppliers' requirements to be provided.
5. Coordinate with the mechanical engineer to provide sump pit and sump pump or connecting the lift pit to the nearest sump pump to prevent water accumulation in the lift pit.

6. Firemen switch should be provided that will enable using one elevator or group of elevators to evacuate people from the building in case of fire.
7. Lift entrance sections, and details should be provided for each floor if they are not typical.
8. Escalators and Travelators pits details with the control panels and equipment installation details should be provided coordinated with other services.
9. Control room coordination: Drawing in details to clarify the equipment's location related to the system.
10. The finishing floor level and wall thickness are should be marked near the elevator entrance for easy reference.
11. Single line diagram with details for intercom system and traffic directory panel, supervisory panel, etc....
12. The final location of the earth leakage relay or earth leakage circuit breaker should be studied and confirmed according to the authority's requirement.
13. A toilet or water pipes passing around the lift shaft are not allowed to prevent pit flooding.
14. Location of alarm bell in Control room to be coordinated with other services.
15. Emergency lights inside the cabin should be considered.

Installations of Lifts, Escalators, and Travelators

the consultant should follow the contractor to start the preparations for the installation of Lifts, Escalators, and Travelators equipment:

1. Coordination meetings between the Main Contractor and suppliers of Lifts, Escalators, and Travelators and the interfaced systems should be held to discuss and agree on the scope of work and all requirements of each supplier of the system (copy from minutes of the meeting to be passed to the consultant for reference).
2. The consultant should instruct the contractor to submit the installation Method statement for the Lifts, Escalators, and Travelators system for the consultant's approval. Installation work shouldn't start before the approval for the method statement or at least approval with notes that can be rectified within installation work, including the following installations' method:

   For Lifts:

   a. Hoistway
   b. Entrance
   c. Machine room work

d. General wiring

e. Car safety installation

f. Counterweight assembly

g. Roping

h. Terminal safety switches

i. Car cage assembly

j. Car door motor

k. Testing and adjustments

For Escalators:

a. Survey and marketing

b. Hoisting

c. Joining and positioning of truss

d. Truss alignment

e. Balustrade and glass panel installation

f. Outer shed thing

g. Wiring and connection

h. Test run, steps, and handrail installation

3. The M.E.P. Contractor should provide all conduits, openings and cables required between (Lift, Esc, and Trav.) and all related and interfaced systems' control panels like B.M.S., Fire Alarm, Access control, Intercom handset in reception or control room, Public Address System Etc.

4. M.E.P. Contractor should provide all conduits, sleeves, and cables required between lifts, escalators, and travelator systems as well as generator/A.T.S.

5. Proper protection for all devices, wires, components, panels ...etc. shall be provided during storing and installation. Lift shaft should be kept clean from water and construction debris.

6. Flow Chart for all activities related to the Lifts, Escalators, and Travelators systems showing the scope of work for each contractor, equipment for the systems to be installed in the control room, interfaced systems, installation requirements, location of power supply, isolator etc.

7. Activity chart and work program for the system, including work execution steps for Civil work, Electrical work, finishing work, testing & commissioning with monthly updates.

8. The contractor and Supplier of the system should provide all engraved labels and identification for all junction boxes, isolators, control panels, interfaces with other systems, emergency push buttons, etc.
9. The supplier should coordinate with the Electrical subcontractor for the required earthing of the system (to be provided as per standard, Authority, and manufacturer requirements).
10. Supplier, Contractor's Site Engineer should daily and continuously monitor the installation activity till completion. Periodical progress report showing site status, requirements ...etc. Shall be submitted by the contractor.
11. Level and finish off all lift walls inside the shaft should be uniform to allow smooth installation of the rails.
12. For Machine less lift:
    a. Shaft ventilation/ cooling should be considered if the lift is exposed to the sun.
    b. The Isolator Location should be approved by the authority (Civil Defense).
    c. Control panel Location and lighting switch to be approved by the Architect Engineer.

13 Break down for all item related to the system shall be proposed by contractor for billing purpose; these broken-down percentages should be agreed by all the parties.

14. The supplier should submit an installation certificate confirming that the installed system is healthy and fit for operation.

Testing and commissioning of Lifts, escalators, and Travelators:

1. Third-party testing of lifts should be done as per local civil defence authorities.
2. Before starting the testing of (L, E and T) systems, the consultant should instruct the contractor to complete the work related to the system and submit a specific T&C Method Statement for testing and commissioning for the consultant's approval.
3. A megger test for all cables and circuit wiring should be completed before starting any termination in the system.
4. Confirmation report from the supplier that the (L, E and T) system installation is completed and ready for operation and testing.
5. Testing Details of Lift, Escalators and Travelators should be as detailed below and to be witnessed by the consultant representative:

General checks and observations:

a. Ensure that the machine room is cleared of all unnecessary objects and is lockable. (Applicable for lift with machine room).

b. Check that Emergency Rescue Instruction is displayed in the machine room or, inside the controller cabinet if the elevator is machine-room-less.

c. Ensure that machine room lighting and ventilation are adequate. (Applicable for lift with machine room)

d. Make sure that the control panel is fixed properly and the connection of wires to the machinery is tight.

e. Check the power supply wires and earthing connections in the main switch and controller to be correct and tight.

f. Check and verify that all other electrical connections are firm and correct.

g. Measure the insulation resistance value using 500 volts Megger, for motor and supply terminals to earth. Note down the values in the Test Report. The value should be more than 0.5 mega ohms.

h. Ensure the continuity of earth circuit from machine room earth terminal to controller, car and landing entrance frame using multi-meter.

i. Check and ensure correct fuse ratings are in the power supply isolator and control panel.

j. Measure the main supply voltage between the three phases using a multi-meter and ensure they are within acceptable limits (as provided by the Authority).

k. Ensure the lift pit is clean and dry.

l. Check and ensure that the pit ladder is installed.

Testing details: Checking safety aspects prior to commissioning:

a. Ensure that the brake is adjusted according to the manufacturer's instructions.

b. Check and ensure that all safety switches are properly connected and that the lift will not move if any of them is operated.

c. Check and ensure that the lift will stop immediately as soon as the landing or car door is opened. (This checks the functioning of the car door contact and the landing door lock contacts).

d. Ensure that the rope fastenings are done as per the relevant instructions.

e. Check that the buffers are installed as per the relevant instructions.

f. Measure the run-by distance of the car and counterweight using measuring tape and ensure that it is as per the Layout Drawing.

g. Check and ensure that all landing doors can be unlocked from outside using the Emergency Key.

h. Ensure that the car cannot be moved upwards electrically when the counterweight is on the fully compressed buffer.

i. Check and ensure that the final limit switches operate before the car or counterweight contacts the buffer.

Function tests:

a. Verify that the buzzer sound can be heard on the main floor/ lobby when pressing the alarm button inside the car. Also, Intercom voice contact should function from the lobby to the car.

b. Verify that the alarm and intercom function during the power supply OFF condition. The light should be on inside the car.

c. Measure the current readings using ammeter in all three phases when the lift is running at rated speed in both directions for no load, half loading and full loaded car. Note down the readings in the Test Report.

d. Measure the leveling accuracy using measuring tape, for empty and fully loaded car in both directions, at one upper floor and one lower floor. It should be within +/-5 mm. Note down the readings in the Test Report.

e. Run the lift continuously for 1hr with full loaded car and check that there are no variations in ride or leveling. Note down in the report.

f. Run the lift downwards from the topmost floor with 100% nominal load at the rated speed. Switch off the power in the lower part of the travel. Verify that the lift car stops fully by the brake.

g. Verify that the brake sustains the static car with 125% load inside at the lowest floor. The lift should not move from the floor level.

h. Verify the safety gear operation when engaging at rated speed.

i. For 110% loading, verify that the buzzer sounds and overload indication being on in the car. The car should not move from the floor and doors should remain open. Car calls should not be registered.

j. Verify the car, landing indications and call registration functions are correct.

k. Check and ensure that the door reopens when there is any object between the closing doors, or when door open button is pushed.

l. Check the Fireman operation function. When the Fire signal/ switch is activated, the lifts should return to the main floor, and doors should open. Landing calls will be inoperative. Only one car call will be in service at a time, and the rest of the lifts will be under the Firemen's control.

m. Check the function of the Emergency Battery Drive (EBD). When the power fails during normal running, the lift car stops immediately within a time delay, the EBD should function automatically. The car should be driven to the next floor level and door should open.

6. Final approval for the system will be subject to.

    a. Successful completion of Testing and Commissioning of the system.

    b. Successful completion of testing by a third party and Civil Defense.

O&M manuals for Lifts, Escalators and Travelators:

1. Description of the project's system.
2. How to use the manual.
3. Troubleshooting guide (for emergency breakdown).
4. Safety instruction at work.
5. Start-up and shutdown procedure.
6. Original catalogues highlighting the material used and the technical specifications.
7. As built drawings including block diagram.
8. List of spare parts.
9. Procedure for routine maintenance for monthly/1 year/5 years.
10. Warranty certificate for 2 years (as per project specifications).
11. Testing and commissioning certificates/reports, including third-party certificates.
12. Emergency information.
13 Manufacturer/supplier information and contact details.
14. Overview and method of operation (sequence and modes of operation).
15. Training for the security, operation, and maintenance team.
16. All guidelines are for using all devices/controllers, etc.
17. System programming and configuration, if applicable.
18. Details of connections and interfaces with other systems and applicable protocols.
19. Details of lift management system operation, maintenance, and drawings details.

Generator and Automatic Transfer Switch (A.T.S.)

Approval for material submittal for Generator and A.T.S.

The consultant should instruct the main contractor to arrange periodical meetings between the Electrical Engineer, the Supplier of the Generator and all other suppliers related or interfaced with the generator, like fire alarm, switch gear, A.T.S., etc.. to discuss and clarify the mode of operation during fire or power supply failure, connection to switchgear, loading, material submittal requirements, work progress, and the new requirements of the Authority and Civil Defense. The material submittal should consider the followings:

1. The system's suppliers for generators and A.T.S. should be listed on the consultant's Approved Manufacturers List.
2. Description of the system (Generator and A.T.S.) and all related components.
3. Compliance statement to be submitted. The supplier should clearly show his compliance to the specifications, non-complied sections to be highlighted with justifications.
4. A.T.S. should be submitted within a separate submittal.
5. The submittal should include the test certificates for the systems.
6. The submittals should include the original catalog for each system's components, highlighting the technical points that exist in project specifications and highlighting the part number of the proposed components.
7. Consultant and Authority approved shop drawings for the emergency load to be sent to the supplier to be familiar with the load and to be included in the submittal.
8. Consultant approval for all mechanical loads connected to the emergency system and fed by the generator to confirm the actual load. The generator should be able to feed 100% loading.
9. Consultant approval for generator room coordination drawings with full details and dimensions including all generator equipment and the A.T.S. (if in the same room) otherwise, the A.T.S. will be shown in different coordination drawing.
10. Operation details and mode of operation approvals for the generator and the A.T.S. describing starting time delay after power failure.
11. The submitted system should include a block diagram showing all interfaced and interlocked systems such as fire alarm, A.T.S., Lifts, and B.M.S. and remote alarm system.
12. Sequence and philosophy of operation should be included in both submittals for generator and A.T.S.
13. Details of the earthing system should be shown for the generator and A.T.S. in both submittals.

14. Identification with engraved labels and ferrules for all cables and control wires should be part of the submittal.
15. Vibration and noise level checks must be part of the submittal.
16. The door size of the generator room must be included in the technical submittal.
17. The fresh air louvers Dimensions and exhaust of the generator room must be part of the submittal. Air flow, heat load calculations and louver sizing should be submitted for consultant's approval.
18. Load test to be carried out at supplier's factories for the generator and A.T.S. before delivering the generator set to the site and to be witnessed by the consultant.
19. The generator set can deliver rated output power at 50C and 90% R.H., non-condensing.
20. Fuel tank (Built in or separate) refer to project's specifications. Sizing calculations should be submitted.
21. Starting batteries should be sealed with maintenance-free lead acid.

For A.T.S.

22. Main Breaker should be 4 Pole.
23 Data required for B.M.S. to be approved by the consultant (monitoring or monitoring and control).
24. A.T.S. Troubleshooting procedure for emergency breakdown should be included in the submittal.
25. Ingress Protection (I.P.) for A.T.S. enclosure should be suitable for the location being installed. The form of construction will follow the project specification.
26. ACB/ MCCB Ratings should be according to the approved schematic diagrams.
27. The A.C.B. should have all protection functions; S.S., O.L., U.V., O.V., E (start/stop, overload, under voltage, over-voltage, earth leakage).
28. All indication lamps must be low voltage with a built-in transformer or L.E.D.
29. Short circuit rating of all circuit breakers should be according to approved drawings and based on the authority regulations or the approved short circuit calculations.
30. The A.T.S. should have a bypass facility (4 poles with 2 manual bypass facilities) and can be operated in auto and manual operations.
31. All MCCBs shall be of electronic type with adjustable settings for overload and short circuits, with the facility to test the tripping using an electrical pulse from a portable unit.
32. All busbars inside A.T.S. should be solid copper bars (multi-strip laminated busbar is not accepted) with a 20°C rise above ambient temperature, i.e., 70°C limiting temperature at full load.
33. Description of communication protocol. The protocol should be compatible with other interfaced systems, and all suppliers should sit in meetings for discussion and final agreement.

Workshop drawings for Generator and A.T.S.

The consultant should instruct the contractor to submit the workshop and coordination drawings for the generator and A.T.S. for the consultant's approval, including the following:

1. Builder's work related to the generator and A.T.S. installations showing all openings, trenches or trays, intake louvre, opening for fuel pipe filling point, cable entries, generator foundation, the opening of exhaust, etc.
2. Control room/ security room coordination drawings showing the location of the Annunciator, monitoring panel and connection to B.M.S. (if any). The remote alarm system location should be decided upon and should be shown in the related coordination drawings.
3. Detailed coordination drawing for generator room with dimensions including the following:
    a. Generator
    b. Routing of exhaust outlet and silencer
    c. Discharge louver
    d. Radiator duct
    e. Control panel/A.T.S.
    f. Fuel tank
    g. Rubber pad (vibration arrestor)
    h. Switch socket outlet
    i. Lighting fixture and emergency light
    j. Ventilation louver
    k. Routing of entry and outgoing cables and control wires.
4. Supplier confirmation for the generator room that the size and height of the generator room comply with the standard (refer to the data of the generator).
5. Details of the earthing system should be shown in the drawings.
6. Discussion with the supplier for the following requirements:
    a. Engine oil drain while changing oil. With oil pumping in and out facility through the electric pump and valve assembly.
    b. Fuel filling point.
    c. Providing partitions in the trench to avoid leaks passing to the cable in case of fuel leakage.

d. Time delay is in case of a high starting load current.

e. Location of A.T.S. panel (air conditioning area).

f. Location of exhaust, ventilation, and air circulation of the generator room.

g. Water tap over the radiator for water filling.

h. The cable connection box should be studied carefully to allow smooth cable connection on the generator, and the A.T.S., especially the A.T.S. cables, will be about 12 cables.

Installation of Generator and A.T.S.

The consultant should follow and instruct the contractor in the installation stage of the Generator, ATS and the interfaced systems' contractors to check and esure the followings:

1. The consultant should instruct the contractor to submit the installation Method statement for the generator and A.T.S. systems for consultant's approval. Installation work shouldn't start before the approval for the method statement or at least approval with notes that can be rectified within installation work.
2. Activity chart for all activities detailing the complete scope of work for each contractor in standby generator and A.T.S. installations clarifying work execution (civil works, electrical interfacing with other systems, finishing work, testing/commissioning and define scope of work with monthly updated).
3. The supplier and Contractor's Site Engineer will continuously monitor the installation activity for both the generator and the A.T.S. Periodical report should include progress and lookahead and should be submitted to the consultant, client and the owner.
4. Break down for all items related to the system shall be proposed by contractor for billing purpose; these broken-down percentages should be agreed by all parties.
5. Flow chart for all system's components, distribution board connected to generator, actual load, size of cable, way of starting, connection to other systems, etc.
6. The contractor should study the site and provide suitable access for loading unloading of the generator and A.T.S. Also, there should be a procedure for future generator replacement or at least a way to replace the major items like the alternator or the silencer by introducing a facility to pull the generator out or provide a suitable crane.
7. The supplier should supervise the installation of the standby generator and should certify that the installation is correct and fit for operation.

Generator Installations

1. Ensure that anti-vibration pads have been installed and have been aligned on the floor at correct locations as per approved drawings.

2. Check the nameplate according to the approved documents.
3. While installing equipment on the mounting pad, ensure that it is aligned with the pad and all relevant and auxiliary systems.
4. After installing the equipment with fixing bolts on the base frame with a pad, ensure it is levelled and aligned after tightening the fixing bolts.
5. The radiator location should be directed to the louvers.
6. Check that the muffler installation is according to the relevant drawings.
7. Check the radiator, it should be filled with the coolant liquid.
8. Check all accessories are fit properly.
9. Check that the storage tank fits well with all installations required.
10. Check that the automatic transfer switch is in accordance with the relevant documents.
11. Check all cables are terminated with the A.T.S. generator & ESMDB.
12. Ensure that the location and alignment of cable entry have been checked for sufficient clearance for entry and glanding of cables as per manufacturers' instructions.
13. Ensure that power cables are terminated at the correct phase bar with the correct landing. An insulation resistance test was conducted prior to termination.
14. Ensure that the auxiliary control panel is installed and connected correctly in accordance with the approved drawings.
15. Ensure that grounding of frame base gen set body, fuel tanks and alternator earthing has been completed.
16. Ensure that equipment is earthed with correct size of conductor and connected to separate earth pit, which has been prepared for the same.
17. Ensure that the battery has been installed at the correct location and that connections are correct as per the approved drawings.
18. Ensure that support frames are securely fastened to the floor.
19. Make sure that there is no leakage from the batteries.
20. Check all connections between batteries are done correctly.
21. Make sure that battery terminals are cleaned, tightened, and fitted with a protection cap.
22. Check that all warning signs are fit.
23. Ensure that the location of battery chargers, heaters, etc., as well as installation and connections, are correct.

24. Ensure that the cables were sized corresponding to the D.C. current ratings.
25. Refer to the B.M.S. data point where the generator is to be monitored (ensure that all requirements have been considered).

A.T.S. Installation

1. The final location of A.T.S. should be according to the final approved coordination drawing of the main electrical room with clear and enough dimensions allowing easy access to the generator room for the equipment installation. The distances between equipment and panels should allow the panel door to be opened, not less than 90 degrees, for easy maintenance.
2. Ensure that A.T.S. is free from any mechanical damage and clean from any dust.
3. Any cutting in the A.T.S. enclosure should be painted against rust.
4. A preliminary insulation test for all cables should be done before glanding or cable dressing.
5. Ensure that the glands' earthing connection is made and conductor continuity is verified.
6. Ensure that all electrical connections are correct and tightened robustly for adequate torque and marked. All shrouds are intact on all live wiring terminals.
7. Check all control wiring are installed correctly.
8. Ensure that all control interfaces are installed correctly (B.M.S. and F.A.S.).
9. Check and ensure all components are labelled and identification markers, ferrules, and warning sign stickers are available.
10. Check the rating of circuit breaker, contactor, size of the cable and bus bar are as per final approved workshop drawings of the system.
11. Check if the phase sequence and colour for cables and bus bar are correct with proper identification.
12. Ensure that circuit breaker mechanical interlocks are checked (circuit breaker is verified with respect to panel door interlock).
13. Ensure that the mechanical operation of the circuit breaker is satisfactory (on/off/trip).

Testing and commissioning of Generator and A.T.S.:

1. Before starting testing of the generator and A.T.S., instruct the Contractor to submit the testing and commissioning method statement for the consultant's approval. The method statement should include the checklist for testing each component with the successful test criteria:

   Generator test:

2. Mechanical Checklist:

a. Check the generator's entire exhaust system.

b. Check inlet louvers.

c. Check outlet louvers.

d. Ensure that there is no leak in cooling water system, lube oil system, fuel line and fumes disposal line.

e. Ensure that fuel system has been checked as per flow diagram and its inlet lines, outlet lines, level and height of day tank, tank drain facility, fuel tank vent and limit switches.

3. Checklist for Engine:

    a. Check engine water level.

    b. Check the oil level.

    c. Check purges fuel line.

    d. Check the correct "Zone" error for the gauge and meters.

    e. Check the belt tightness of the radiator fan.

    f. Fan to guard clearance.

    g. Settings of thermostat

4. Cooling water system:

    a. Check the water level.

    b. Check control for low coolant level sensor operation, high-temperature alarm and shutdown.

5. Lube oil System:

    a. Check the oil level and ensure that there is no leak from the sump or drain.

    b. Check the lube oil temperature high/low alarm and shut it down.

6. Ensure that all power and control cables have been installed and connected as per approved drawings.

7. Ensure that the stator wiring of generator insulation resistance and winding resistance tests have been carried out with suitable precaution (i.e., isolation from another part of the generator, presence of safety personnel).

8. Check ambient temperature and winding temperature for various load conditions and record it (i.e., at "Normal" and "Standby" load conditions).

9. Ensure that the following are checked on the Generator Control Panel (desk) and completed:

    a. Confirm termination and identification of control wires.

b. Check the 'Zero' error of measuring instruments (with valid calibration and in line with the meter manufacturer's manual/recommendations).

c. Check the polarity of D.C. voltage.

d. Check various types of settings for protective relays and breakers (based on approved coordination study).

e. Check and adjust timer setting relay as per manufacturer's recommendations/based on approved coordination study.

f. Check output voltage.

g. Check the frequency.

10. Check battery condition & terminals, battery voltage, relevant settings on control panel, AC/DC alarm and battery charger.
11. Check the recommended parameter of the electronic governor on the control panel.
12. Confirm the polarity on the digital voltage regulator's terminals and check another parameter as per the manufacturer's recommendations.
13. Ensure that the charger has voltage alarm to indicate visually or audibly the battery terminal voltage when a dangerous level has been reached.
14. Check all control wiring from the day tank float to the fuel pump.
15. Test the interconnecting cables from the generator to the automatic transfer switch by Megger.
16. Check the continuity of the protective conductor.
17. Earth resistance test.
18. Hipot (high potential) test.
19. Earth continuity test to verify cable armouring and metal sheet bonding to earth.
20. Phase rotation test and phase correspondence test.
21. Check all input-output points to B.M.S. during generator operation.
22. Final approval is subject to the successful testing and commissioning of the generator.

A.T.S. test:

1. Before starting A.T.S. testing, instruct the Contractor to complete the testing and commissioning of related systems such as generator, MDB and, SMDB, etc.
2. Certificate from the supplier of A.T.S. after completing the installation of A.T.S. by the Electrical Contractor that the system is ready for testing and commissioning.

3. Visual test: Ensure that all ratings of circuit breaker/Contractors/fuses … etc., are as per the approved workshop drawing. Operation (mechanically on/off/trip)/functions are correct.
4. Megger test for A.T.S. and for cables between generator A.T.S. and MDB to be done.
5. Operational test:
    a. Ensure that the system is working according to the approved mode of operation.
    b. Ensure that the functions of all indicators, selector switches, voltmeters, ammeters, and other accessories are correct.
    c. Check that the anti-condensation heater is functioning properly.
6. Final approval is subject to the successful testing and commissioning of A.T.S.

Operation and Maintenance Manual for generator and A.T.S.

1. Description of the project system.
2. How to use the manual.
3. Troubleshooting guide (for emergency breakdown).
4. Safety instruction at work.
5. Start-up and shutdown procedure.
6. Original catalogues highlight the material used with part numbers.
7. As built drawing including schematic riser diagram/block diagram.
8. List of spare parts.
9. Procedure for routine maintenance for monthly/1 year/5 years.
10. Warranty certificate for 2 years (as per project specifications).
11. Testing and commissioning certificates/reports.
12. Emergency information.
13. Manufacturer/supplier information and contact details.
14. Overview and method of operation (sequence of operation).
15. Training for the security, operation, and maintenance team.
16. All guidelines for using all devices/controllers, etc.
17. System programming, configurations, and communication with other systems.

Earthing and lightening Protection

Approval for material submittal for Earthing and Lightening Protection Systems.

The consultant should instruct the contractor to arrange periodical meetings between the main contractor, Electrical engineer & supplier of the Earthing and Lightening protection systems to provide the following:

1. Discussion to clarify installation requirements as follows:
    1. Water level inside the project plot.
    2. Soil resistivity report.
    3. Provision of well, ring or mesh for earthing cables.
    4. Location of test points.
    5. The earth rod pit is located inside the basement or outside the building.
    6. Down conductor (surface/ inside sleeve or part from re-enforcement on the roof or any metal cladding around the building.
2. The system supplier should be listed on the consultant's approved manufacturers list.
3. Description of earthing and lightning protection systems and all related components.
4. Compliance statement to be submitted. The supplier should clearly show his compliance to the specifications, non-complied sections to be also highlighted with justifications.
5. The submittal should include the test certificates for the systems.
6. The submittal should include the original catalog for each component of the system, highlighting the technical points that exist in project specifications and highlighting the part number of the proposed components.
7. The supplier should submit a sample board containing all components to be used for consultant approval.
8. Technical calculations for the Lightening Protection System should be submitted by the supplier specialist for the following:
    a. Lightning protection level (L.P.L.).
    b. Lightning Protection System (L.P.S.).
    c. Number of down conductors.
    d. Earth resistance Test. The test result will be the guide to decide if well, ring or mesh is required for the earthing cables connection to ground.

9. Calculation from supplier to be part of the submittal to decide the details of all earthing system required by all authorities (Power supply, Transformer, Tel, GSM, Generator. etc.)
10. All earth pits inside the building should be watertight to prevent water leakage to the building due to water table.

Shop drawings for Earthing and Lightening Protection Systems:

The consultant should instruct the contractor to prepare the workshop and coordination drawings for the Earthing and Lightening Protection Systems to be submitted for the consultant's approval with the following requirements:

1. The location of the earth pits will be determined after coordination with all other services. (Avoid entrances and front of house areas).
2. Location of earth rods for pits inside the building (basement) in coordination with all other services around.
3. Equipotential bonding conductors for all cladding around the building, mechanical equipment on the roof, T.V. antenna and all consumer wiring installations
4. Location of test points for earth pits.
5. Routing of bars on the roof.
6. Location of air terminals on the roof.
7. Location of down Conductors (surface/inside the sleeve/part from re-enforcement steel bars).
8. All connections with metal equipment and installations on the roof.
9. All connection with metal cladding around the building taking into consideration if the insulation material is used in between or not.
10. Location and level of all sleeves between buildings and outside, taking into consideration water level and waterproofing around the building.
11. The water level should be identified and marked on the drawing.
12. The type of soil investigation should be marked on the drawing to decide the type of material for the earth rod (copper or stainless steel).
13. Location of earth pit (If basement covering the whole plot, then the pits can be in the basement). The earth pit should be of watertight in the basement.
14. The decision whether down conductors used are surface type or separate re-enforcement steel bars within the columns and extended by welding to the roof.

15. Details of earthing system for low current systems.
16. All earth pits inside the building to be watertight.

Installations of Earthing and Lightening Protection Systems.

The following activities should be checked during and after the installation of Earthing and Lightening Protection systems.

1. The contractor should submit the Installation Method Statement for the consultant's approval.
2. Earthing conductors' sizes should be as shown in the approved shop drawings and specifications.
3. The cross-section area of the equipotential bonding conductor should be selected as specified and as per Local Authorities Regulations.
4. Check and ensure the copper conductor/ tape is fixed with saddles.
5. Ensure that all bolts are tightened.
6. Ensure that the earth rod is of the approved size and type and installed in the approved location as per the approved coordination drawing.
7. Ensure that the earth rod of 16mm/ 20mm dia. and 1200mm length is driven to the ground with approved accessories like (stud, coupling… etc.).
8. Ensure the concrete inspection pit with a cover according to the approved coordination drawing is provided with concrete around it for proper support in case located in an area with no interlock and named as follows:
    a. For lightning protection systems (lightning protection),
    b. For electrical system (electrical earth).
    c. For low current system (low current earth).
9. Ensure type G clamps are used on the earth rod for termination with the earth conductor.
10. Ensure type G clamps are used on Earth Road for termination of the neutral conductor coming from the transformer star point (network power supply).
11. Ensure that the transformer body earthing is terminated on the main earth bar installed in the MDB room.
12. Ensure that the earth bar of the approved size is installed in the proper location in the electrical room.
13. Ensure that the earth conductor for MDB is connected to the earth bar.
14. Ensure that the earth continuity conductor for the incoming/ outgoing cables and the earth ring of the cable glands are terminated on the earth bar.

15. Grease should be provided on earth rods after testing and commissioning of the system.
16. The earth pit manhole should be checked and coordinated with the surrounding FFL.
17. Separate earth rods and separate earth pits should be used for D.C. current, low current systems and A.C. current.
18. All earthing system for authorities such as Telecommunication, Transformer, GSM, Generator, Main Electric Room (MDB) to be segregated from lightening protection earthing pits.
19. A flow chart for all activities describing the complete scope of work to be submitted shows system components, the number of earthing cables on the facade of the building, the number and type of earth pits, and the type of rod.
20. Identification with engraved labels and ferrules to be provided for the system components.
21. The earth pit located inside the basement should be of double-flange earth and sealed before casting the basement slab.
22. If the design of the lightning protection system is made to use a dedicated steel bar inside the columns and to be welded continuously from the roof to the basement and to the connection to the earth pit, then this steel bar should be painted to be recognized as the lightning protection bar. **See the details In 8-1:**

8-1 Dedicated steel bar for lightening protection system, this bar should be welded to be continuous from top to the connection with earth pit, and to be painted to be recognized from other bars after pouring the concrete.

Testing and commissioning of Earthing System.

Before starting the testing of the Earthing System, the consultant should instruct the Contractor to provide and submit a testing and commissioning method statement for the consultant's approval. The method statement should include the check list for testing of each component with the successful test criteria.

The followings are the tests that will be carried out on earthing and bonding. Taking into consideration, completion of installation and isolation of all power supply during the test.

a. Visual Inspection:

i. All earth electrodes in earth pits should be visually inspected for proper installation and connection.

ii. All protective earth conductors from the earth electrode to the MDB and distribution boards should be visually inspected for proper installation and connection.

iii. Check that all materials are according to the approved submittal and approved workshop drawings.

iv. The distance between two electrodes in the ground should be not less than two-thirds (2/3) the length of the first electrode.

v. Check if the generator neutral is grounded.

vi. Check all metallic conduit, mechanical equipment, water heater, lighting fixtures, cable tray… etc. are bonded by earth terminal (size of grounding conductor should comply to Authority regulation).

b. Measurement of earth electrode resistance

The earth electrode resistance of each earth pit should be measured with bridge megger after removing the protective conductor connection. The reading should be recorded and verified with the permissible limit (1 ohm for earthing system pit).

c. Continuity Test:

i. The continuity resistance of protective conductors should be measured with a D.C. ohm meter, values should be be recorded and verified to be within the permissible units to ensure proper earthing.

ii. The continuity resistance between the main earth terminal at the distribution board and the metallic frameworks of the farthest equipment should be measured via a D.C. ohm meter, values should be recorded and verified with the permissible limit to ensure proper bonding.

Testing and commissioning of Lightening Protection System:

Before starting the testing of the Lightening Protection System, the consultant should instruct the Contractor to provide and submit a testing and commissioning method statement for the consultant's approval. The method statement should include the check list for testing of each component with the successful test criteria.

1. The followings are the tests that will be carried out on Lightening Protection System. Taking into consideration, completion of installation and isolation of all power supply during the test.

1. **Visual Inspection:**

    1. The lightening protection system installation should be visually inspected to ensure proper joints of copper tapes with accessories and earth electrodes in the earth pits.
    2. All joints and bolts should be cleaned and tightened to maintain good continuity resistance.
    3. Ensure equipotential bonding conductors' cross-section area for cladding around the building, mechanical equipment, T.V. antenna... etc., are selected as specified.
    4. Check that all materials are installed as per approved workshop drawings.
    5. The distance between two electrodes in the ground should be not less than two-thirds (2/3) the length of the first electrode.
    6. Check that all down conductors are fixed with proper clamps at proper intervals without any sharp bends.
    7. Check that all air terminals are installed and connected to the lightning protection network.
    8. For the lightning protection system using reinforcement steel bars instead of copper strips; check that all steel bars for the lightning protection system are continuous from the earth rod up to the roof).

2. **Measurement of earth electrode resistance**

    The earth electrode resistance of each earth pit should be measured with a bridge megger after removing the protective conductor connection, and the reading should be recorded and verified with the permissible limit. (10 Ohms for lightning Protection pit).

    1. The continuity resistance of protective conductors should be measured with a D.C. ohm meter, and values should be recorded and verified to be within the permissible units to ensure proper earthing.

O&M Manuals for Earthing and Lightening Protection Systems.

1. Description of the project systems.
2. How to use the manual.
3. Safety instructions at work.
4. An original catalogue highlighting the used material with part number.
5. As-built drawings.
6. List of spare parts.
7. Procedure for routine maintenance.

8. Testing and commissioning certificates/reports.
9. Emergency information.
10. Manufacturer/supplier information and contact details.

Lighting Fixture and Lighting Control System

Approval for Material Submittal for Lighting Fixtures and Lighting Control System:

The consultant should instruct the contractor to arrange periodical meetings between the Contractor's Electrical engineer, the suppliers of Lighting fixtures and Lighting control system and any other related or interfaced system such as B.M.S., Dimming and theatre Systems to discuss and clarify the material submittal requirements, installation requirements, rating, coordination with other work, and work progress, etc. The contractor should provide the followings for the consultant's approval:

Light fixture submittal requirements:

1. The fittings' suppliers are listed in the consultant's Approved Manufacturers List.
2. Description of the lighting fixtures and all related components, including the override switches.
3. Compliance statement to be submitted. The supplier should clearly show his compliance to the specifications, non-complied sections to be highlighted with justifications.
4. The submittal should include the test certificates.
5. The submittal should include the original catalog for each fixture highlighting the technical points that exists in project specifications and the part number of the proposed fittings.
6. The contractor should submit samples of the lighting fixtures for approval. The consultant and the client should approve the required colour temperature for the lights in all locations.
7. The minimum cross-section of the wiring should be 0.75 sq mm, heat resistant at 105° C and should have specific terminals inside fitting for wire connection.
8. All light fittings should be L.E.D. type with suitable drivers for use. The supplier should advise about the maximum number of fittings for each driver and the maximum distance between the driver and the fittings.
9. For dimming systems, coordination between suppliers to ensure compatibility of the submitted fittings with the dimming system.
10. All electrical components should be assembled and installed in the manufacturer's factory; local assembly for electrical components will not be accepted.
11. Luminaires colour is to be coordinated with Architect and to be approved by the client. The client should select the required colour and sign the selected fitting on the submitted sample.
12. The depth of the fitting is to should be coordinated with all services above the false ceiling and in wall claddings (if applicable).

13. Ingress protection (I.P.) for the selected fitting will follow the design and the location of installation.
14. Lux level calculations to be provided for several areas to select the quantity and the type of the light fittings.
15. All required installation accessories should be included in the light fittings.
16. A cutout/ fuse should be provided in the light fitting.
17. Light fittings to be selected based on the installation location of fixtures (false ceiling, surface, bracket, above door Etc.)
18. Downlight ventilation should be from the bottom or side (follow the Authority regulations).
19. Terminal box for recessed and wall mounted fixtures should be provided in suitable place to be hidden. For outdoor and roof fittings, connection of conduit to the terminal box should be form the side or from top to prevent water penetration to the P.V.C. conduit.
20. Fitting accessories like fittings with reduced voltage transformer should contain all accessories including the transformer inside the fitting and earthing terminal.
21. All downlights inside the bathroom should have safety protection glass.
22. All external fixtures shall have stainless steel screws.
23. The diffuser should withstand the ultraviolet rays and its color should be stable and won't change to yellow by time.

Lighting control System submittal requirements:

1. The system's supplier is listed in the consultant's Approved Manufacturers List.
2. Compliance Statement is provided according to the project's specifications. The supplier should clearly show his compliance to the specifications, non-complied sections to be also highlighted with justifications.
3. Description, mode, and philosophy of operation of the lighting control system and all related components.
4. The submittal should include the original catalogue for the Lighting Control System and all related accessories (sensors and override switches), highlighting the technical points that exist in project specifications and the part number of the proposed fittings.
5. The submittal should include the test certificates.
6. A flow chart shows all related activities and identifies interfaced systems to the light control system.
7. A block diagram/ riser diagram should be included in the submittal to clarify all interconnections with all other systems, such as B.M.S., dimming, outside lighting, theatre lighting, etc.

8. Identification, engraved labelling, and ferrules to be part of the submittal.
9. A list for all equipment related to the lighting control system to be placed inside the control, B.M.S. room clarifying the dimensions to prepare coordination drawing.
10. Details of the earthing system should be included in the submittal.
11. The submittal should include the required calculations to decide the quantities of used heat and motion detectors that will ensure continuous motion detection without dead zones.
12. For the staircase, a multifunction timer contactor should be used; the staircase light can be switched on and off from all floors.
13. All used wiring and cables in the lighting control system should be submitted with the submittal for consultant approval.
14. Communication protocol with other systems, such as B.M.S., should be studied and clarified (preferring compatible protocols).
15. The lighting control programming procedure should be submitted with the submittal.
16. The troubleshooting procedure for the lighting control system breakdown should be part of the control panel submittal.

Workshop drawings for lighting fixtures and lighting control systems.

The consultant should instruct the contractor to prepare the workshop and coordination drawings for the lighting and lighting control system (with lighting drawings) to be submitted for the consultant's approval. The followings should be considered in the workshop drawings:

1. Workshop drawings should show location of motion detectors, override switches, lighting control panel, power supply, etc.
2. Identification with ferrules for all wires between the power supply, control panel, and sensors.
3. All power supplies which are required for operation and system control panel should be from emergency supply (based on project specifications) and fed by separate section (breaker and ELCB).
4. All override switches should be near the lighting control panel.

Installation of lighting fixture and lighting control system.

The consultant should follow the contractor to start the preparations for the installation of light fittings and Lighting control system.

The notes shown below should be taken into consideration in the installation of the lighting fixtures and lighting control system:

1. The consultant should instruct the contractor to submit the Installation Method statement for the light fittings and Lighting Control System for consultant approval. Installation work shouldn't start before the approval for the method statement or at least approval with notes that can be rectified within installation work.
2. Ensure the consultant's approval for the workshop and coordination drawings for the lighting fixture and lighting control system.
3. When the procured material reaches the site, site engineer should check the received materials ensuring materials are matching the approved submittal. Supplier should provide and to submit relevant factory test reports.
4. The contractor should ensure full protection for the material inside the stores.
5. The contractor should protect the light fittings and the Lighting Control System when being installed and to be protected from being hit due to other construction activities. The area near and around the installed material should be be kept clean form other services and to be protected from water.
6. Supplier, Contractor's Site Engineer should daily and continuously monitor the installation activity till completion. Periodical progress report showing site status, requirements ...etc. Shall be submitted by the contractor. The supplier engineer should certify the installation, confirming it is correct and fit for energization at the end of installation.
7. Break down for all item related to the system shall be proposed by contractor for billing purpose; these broken-down percentages should be agreed by all the parties.
8. The contractor should provide the engraved labelling and identifications for all system equipment and accessories, wirings, Etc.
9. Earthing for the system should be provided according to the manufacturer's and authority's recommendations.

O&M manuals for light fittings and Lighting Control System.

1. Description of the project system.
2. How to use the manual.
3. Troubleshooting guide (for emergency breakdown).
4. Safety instruction at work.
5. Start-up and shutdown procedure.
6. An original catalogue highlighting the used material with part number.

7. As built drawing including schematic riser diagram/ block diagrams.
8. List of spare parts.
9. Procedure for routine maintenance for monthly/1 year/5 years.
10. Warranty certificate for 2 years (as per project specifications).
11. Testing and commissioning certificates/reports.
12. Emergency information.
13. Manufacturer/supplier information and contact details.
14. Overview and method of operation (sequence of operation).
15. Training for the security, operation, and maintenance team.
16. All guidelines for using all devices, controllers, sensors, etc.
17. System programming and configuration.

Chapter 9

Installations (Part 4) Low Current Systems

This chapter discusses the Low Current Systems, including the following:

1. Fire Alarm System.
2. Emergency Lighting System.
3. Access Control System.
4. Intercom System.
5. CCTV System.
6. Traffic Barrier System.
7. Public Address System.
8. MATV System.
9. Parking Management System.
10. Uninterruptible Power Supply System (UPS).
11. Nurse Calling System.
12. Central Clock and Bell System.
13. Private Branch Exchange System. (PABX).
14. Smart Lighting and Dimming Control System.
15. Guest Room Management System. (GRMS).

Fire Alarm Systems

Approval of materials Submittal for Fire Alarm System

The consultant should follow the contractor to arrange periodical meetings between the Electrical engineer and the suppliers of fire alarm systems and related equipment to discuss and clarify the mode of operation during a fire (cause and effect matrix), new requirements of Civil Defense, interlocks and interfaces with other systems, material submittal and work progress. The contractor should provide the following for the consultant's approval for the material submittal:

1. The selected brand or manufacturer is listed in the consultant's approved manufacturer list.
2. A brief description of the fire alarm system and sequence of operations is to be submitted.
3. Compliance statement to be submitted. The supplier should clearly show his compliance to the specifications, non-complied sections to be highlighted with justifications.
4. The original catalogue highlights all technical points in the specifications and the components used.
5. Detectors to be selected, and the system is to be designed based on the Authority regulation. The multi-sensor in the open kitchen, heat in the closed kitchen, beam detectors in wide workshops…etc.
6. A fire alarm system should be provided for false ceilings of more than 80 cm in height and in trenches.
7. Indication lamp on control panel should indicate the provision of the power supply. Power supply should be from emergency DB if applicable.
8. The control panel should be capable of interfacing with all related systems of the project according to the design, specifications, and Authority regulations.
9. Each fire alarm zone or loop should have 25 % spare for future added devices.
10. Submission of sample board.
11. Valid supplier trade license with expiry date, and ministry of trade certification as a legal supplier of the material in the submittal.
12. Flow chart for all Electro-mechanical equipment interfaced with fire alarm system according to the specification and Civil Defense requirements, including the description, type of the signal, Alarms, and indications on FACP.
13. Block Diagram and Riser Diagram for fire alarm system shall include details for all connections/ interfacing with all Electro-mechanical equipment/devices, Power supplies required for the fire alarm system to be mentioned.
14. Each loop in the fire Alarm panel should have spares for future addition, with remote indicator for each closed area.

15. The submittal should include the requirements for connection to Auto Dialer or Messaging system (RMS).
16. The cause and effect matrix for all systems (if applicable), including electromechanical equipment, clarifies the method of operation or sequence of operation in detail as per the specification and requirement of Civil Defense. The evacuation procedure for all causes shall be submitted in chronological arrangement.
17. Sound-level calculations based on civil defence requirements should be submitted for approval. (Inside flats, the sound level should be 15-20 dB for sounders behind closed doors.
18. Fire Alarm cables should be submitted for approval as a part of the fire alarm system.
19. Communication compatibility with BMS and all interfaced systems.

Workshop drawing for Fire Alarm Systems

The consultant should follow the contractor to submit the workshop drawings for Authority (Civil Defense) and then for consultant approvals. Shop drawings should include the followings:

1. Workshop drawings for fire alarm system should include all Electromechanical equipment which are interfaced with fire alarm system like mechanical equipment, access door, gate barrier, firefighting pump, lift, dampers...etc.
2. Fire Alarm system to be provided for false ceilings with more than 80 cm height and in trenches.
3. The control room/ Security room drawing should include the location of the main equipment for the fire alarm system with furniture layout, 24x7 Auto dialer panel and connection to the other systems like BMS ... etc.
4. All junction boxes are required or dedicated for connection with other interfaced systems.
5. The location of all installations of fire alarm components on and above the false ceiling should be coordinated with all other systems (minimum clearance of 30 cm).
6. All interfacing units should be located in an accessible area; installations for interfacing units above false ceilings are not accepted.
7. Location of all Electro-mechanical systems that are connected to the Fire Alarm System.
8. All comments by Civil Defense on the design drawing should be considered and implemented.
9. The location of sounders (inside and outside the building) and controllers should be according to Civil Defense regulations and recommendations.
10. Zoning of bells should be arranged and distributed between circuits or risers, so in a stair or location that has a malfunctioned bell, the other stairs will function normally as it is fed from different circuits.

11. The fire alarm system should be fed by the emergency DB (if applicable). Separate sections (ELCB and MCB) using fuse connecting unit with indicating lamp should be used.
12. Earthing Details for the Fire Alarm system should be shown in the drawings and according to manufacturer requirements.
13. The location of the main fire alarm control panel and sub-fire alarm control panel should be shown with cable routings and communication protocols. Location to be coordinated with other services and plaster or finishing.
14. Fire Jack telephone to be provided in or near the staircase according to the Authority Regulation.
15. Show the Network and the communication protocol between control panels, interfaced systems and components.
16. Show location and wiring for Civil Defense 24x7 auto dialer panels and other related systems (power and data wires).
17. Mounting height of all devices and accessories should be shown clearly on workshop drawing or to be approved in schedule of heights.
18. Shop drawings to be signed by the suppliers of the material. The signature give indication that the supplier specialist is accepting the installation and the installation details is matching the supplied system.
19. Connection details in the control panel (terminal block) for all incoming and outgoing cables and wires should be provided.
20. Workshop drawings should be submitted for Authority (Civil Defense) approval after being approved by the consultant. Fees should be paid by the contractor or the client according to the project scope.

    Note: Break down for all item related to the system shall be proposed by contractor for billing purpose; these broken-down percentages should be agreed by all the parties.

Installations of Fire Alarm systems

The consultant should follow the contractor in the installation stage of the Fire Alarm System to check and ensure the following:

1. Submission of method statement for installation activities for the whole system with all related interfaces and components. Suppliers should ensure approval of shop drawings before starting installation activities.
2. Completion of Conduit and containment installation for fire alarm wires connecting all devices, interfaced systems, sub, and main control panels.

3. Fire alarm wiring should be installed in trunking, GI conduits, or to be fixed on the ceiling by fire rated strips from the same manufacturer.
4. Remote indicators shall be provided for all closed-area devices.
5. Fire alarm devices or components to be coordinated with all other systems in service areas like mechanical room, Electrical room, parking Etc.
6. Proper protection for all devices, wires, junction boxes…etc. Shall be provided during installation.
7. The mounting height of devices should be as per approved workshop drawings and the height schedule.
8. Break glass should be located directly under the sounders (both on one line).
9. Identification for the system components by using engraved labels and ferrules (devices, wiring…etc.) should be provided.
10. Indication lamp on control panel should indicate the provision of the power supply. Power supply should be from emergency DB if applicable.
11. Covers should be fixed for each fire alarm point for protection till the testing and commissioning stage.
12. Coordinate with suppliers regarding the type of the required control signal for each interfacing with other systems like (volt-free contact, protocols, digital or analogue signals…etc.).
13. Fire alarm cables should be selected as per the project specification and the Authority regulations, taking into consideration voltage drop:
    a. Fire resistance cables for life safety (FP200) should be used for all circuits between connections and interfaces.
    b. All power supply cables for electromechanical equipment connecting to the fire alarm system and firefighting should follow the design circular related to the Authority. Terminal block connection details inside the control panel and identification for all in and out cables/wires shall be provided.
    c. For electrical substations and oil fields, the wiring should be done usually by using MICC cables (follow the project specifications).
14. All wires should be terminated properly to the correct terminal for all devices, control panels, interfaces, etc.…
15. The supplier and Contractor's Site Engineer should monitor daily and continuously the installation activity till completion. Periodical progress report showing site status, requirements ...etc. Shall be submitted to the consultant by the contractor.
16. All control wires and power cables shall be with extra length for connection to the equipment's terminal.

17. Ensure all field devices have been installed on the correct back boxes.
18. Ensure all interconnections between the Fire Alarm Panel and lift, BMS, AHU, GAS, Etc. are completed and tightened according to the approved shop drawings.
19. Ensure all panels, detectors, sounders, flashers, and break glass switches are installed correctly according to the approved shop drawings.
20. Ensure that addressable devices are correctly installed and all logic addresses are recorded for T&C.
21. Activity chart, work progress and look ahead program of work for Fire alarm system clarifying work execution (civil works, electrical interfacing with other systems, finishing work, testing/commissioning and define scope of work with monthly updated).

Systems and equipment interfaced with the Fire Alarm system.

1. Electrical equipment (systems) connected to the fire alarm system:
    a. Emergency lighting system.
    b. MDB (switchgear).
    c. Lift and escalators.
    d. Gate barrier.
    e. Access control system (Emergency doors / sliding doors).
    f. 24x7 Civil Defense Panel.
    g. GPRS.
    h. Building Management System.
    i. Public address system/sound system.
    j. CCTV system.
    k. Emergency exits and Sliding doors
2. Mechanical equipment (systems) connected to the fire alarm system:
    a. FM 200 system is for the MDF room and the main electric room.
    b. Foam system for generator.
    c. Smoke management system for staircase/corridor/lift lobby (Pressurized fan/Fresh air handling unit/SEF).
    d. Jet fan system (Exhaust fans)
    e. Building Management System (BMS).

f. Firefighting system (Sprinkler system/fire hose reels).

g. Fresh air handling unit.

h. CO2 system.

i. Air handling unit.

j. Smoke exhaust fans.

k. Gas system.

l. Fresh air fan.

m. Aspirating system.

n. Garbage chute.

o. Fire and Jockey pumps.

p. Fire Diesel generator

q. Electrical zone dampers.

Testing and commissioning for Fire Alarm System

1. Before starting the testing of the fire alarm system, the consultant should instruct the contractor to complete all work related to the system. The contractor should submit a specific method statement for testing and commissioning, including a checklist showing the standard or successful readings and status in the submitted format.
2. Confirmation report from supplier specialist that the Fire Alarm System installation including all other interfaced systems are completed and are ready for operation and testing.
3. Megger test (loop resistance test) for all circuits (wiring) before starting the termination of the system's components.
4. Visual Checking for the system ensures the following:
    a) All installations of fire resistance cables should have proper laying and termination with Master and repeater FACP, battery, and all other devices.
    b) Check for battery condition and operation by simulating power failure.
    c) The main control panel, heat detector ...etc. are identified and labelled by using engraved labels and ferrules.
    d) Mechanical checking for all components inside the control panel.

e) Before energizing, ensure completion of the control panel programming for all field devices with their respective logic address.

5. Function Test:

   a) This should be carried out for all systems interfaced with the fire alarm system (FAHU, lift ... etc.).

   b) Testing for each device by simulating a fire condition by generating smoke, heat, breaking glass ... etc., for checking the following:

      i. Sound level, audibility, and clarity of announcement.

      ii. Display of detector number

      iii. LED indication on the detector

      iv. Mimic panel

      v. Sounder circuit

   c) All components have been installed according to the approved material submittal and approved shop/ coordination drawings.

   d) All remote LED indicators are working satisfactorily in normal and alarm conditions.

6. Simulate a false alarm and check the following responses:

   a) Lift: Should come down to the ground floor for firefighting team usage.

   b) Door: All doors are to be released, and emergency doors should be operable.

   c) Staircase Exhaust fans: To work at high speed to increase the air pressure in the staircase.

   d) FAHU: To be in top condition.

   e) BMS: Receiving the signals from activated components.

   f) Civil defence/ Auto dialer: functioning.

   g) The event is being recorded in the control panel memory; the event can be printed.

7. Final approval for the system is subject to the following:

   a) Successful testing and commissioning of the system.

   b) Approval of Civil Defense.

Civil Defense requirements for Fire Alarm System handing over

The consultant should follow the contractor to provide the following for the successful handing over of the Fire Alarm system:

A. Approved workshop drawings.

B. Overview and mode of operation (sequence of operation) for fire alarm system.

C. Brief description of the system's performance.

D. Evacuation procedure.

E. Letter to the authority (Civil Defense) requesting for site inspection.

F. Confirmation letter from supplier (certificate) confirming work completion.

G. Completion of room identifications for all rooms and areas related to the system ex (Electrical room, pump room, garbage room ... etc.).

H. Paying the Civil Defense fees required for inspection and certification.

I. Undertaking letter (if required) to finish the work or to complete civil defence comments.

J. Completing the Civil Defense requirement for connection to 24x7 panel that includes:

   1. Provision PVC or GI Trunking between the 24x7 panel and fire alarm panel for power supply and communication connections.

   2. Provision for PVC or GI conduit for antenna.

   3. Completing the following requirements for the Fire Alarm System:

      Low Rise Building

      a. One telephone/data cable (4 cores).

      b. One power cable 1.5mm^2.

      High Rise Building

      a. Three CAT5 data cables, each one (8 cores).

      b. One power cable 1.5mm^2.

   4. Completing the following requirements for the Fire Fighting System:

      a. Two CAT5 cables of (8 cores) each to the firefighting control panel.

      b. (8 cores) data cable CAT5 to the water tank.

      c. 1.5mm^2 power cable (one only).

   5. Elevator System:

      a. One (8 cores) CAT5data cable for each elevator.

      b. 1.5mm^2 power cable (one only).

Note: the above cable requirement may change based on manufacturer or Authority requirements or due to the requirement for the latest technology.

Operation and maintenance manuals (O&M) for fire alarm systems.

The consultant should follow the contractor to provide and submit the operation and maintenance manuals for approval by the consultant, the manuals should contain the following:

21. Description of the project system.
22. Clarifications about using the manual.
23. Description of the Civil Defense panel (24x7) and the connection with the fire alarm panel.
24. Description of the interfaced systems with the fire alarm control panel.
25. The cause-and-effect matrix.
26. Troubleshooting guide for the system.
27. Safety instructions and requirements at work.
28. Start-up and shutdown procedure.
29. An original catalogue highlighting the used material.
30. The as built drawing includes a schematic riser diagram showing the interfacing of the fire alarm system with other systems.
31. List of spare parts.
32. Procedure for routine maintenance for monthly/1 year/5 years...etc.
33. Authority No Objection Certificate (NOC).
34. Warranty certificate for 2 years (as per project specifications).
35. Testing and commissioning certificates/reports.
36. Emergency information.
37. Manufacturer/supplier information and contact details.
38. Overview and method of operation (sequence of operation).
39. Training for security, operation and maintenance team. (It may require the submission of a training manual according to the project specifications).
40. Evacuation procedure for each building.
41. System programming, configurations, settings with the used software details and communication protocol.
42. All guidelines for using devices and controllers inside the control room.

Emergency Lighting System

Material Approval for Emergency Battery System (EBS).

The consultant should instruct and follow the main contractor to arrange periodical meetings between the Electrical Engineer, the Supplier of the emergency lighting system and other related systems suppliers (fire alarm, switch gear, CCTV, etc.) to discuss and clarify the mode of operation during the fire and the latest requirements of Civil Defense. The contractor to be notified to provide the following for the material submittal:

1. The selected brand or manufacturer is listed in the consultant's approved manufacturer list.
2. The martial submittal for emergency lighting system to be submitted for consultant's approval including system description and sequence of operation.
3. Compliance statement to be submitted. The supplier should clearly show his compliance to the specifications, non-complied sections to be highlighted with justifications.
4. The submittal should include the original catalogue highlighting the selected components and the technical points that comply with the project specifications.
5. The material submittal should include at least the riser diagram for the emergency lighting system with full details, including the system's loops or zones, CBU, Emergency lights, Exit lights, connection to BMS and fire alarm system, etc.
6. 10% spare to each circuit (loop or zone) should be added for future added lighting (as per project specifications).
7. Valid supplier trade license with expiry date, and ministry of trade certification as a legal supplier of the material in the submittal.
8. Flow chart and block diagram for major system's parts and components clarifying the location, Power supply details, Lighting circuit reference, Number of emergency lights...etc.
9. The mode of operations is to be clearly explained in the submittal.
10. The panel and the subpanels and components IP rating should be suitable for the installation location. (For wet area in Mechanical room should be IP 65).
11. The submitted material should include proper and approved Identifications and engraved labelling.
12. Lighting Calculations for emergency lighting system should fulfill the required lux level according to the standard (1lux-10lux from below the light till the midpoints between 2 lights) or as required by the project specifications. The calculations to be submitted to the consultant for his approval.
13. Power consumption calculations and battery sizing considering the number of fittings in each circuit, the number of circuits in the system, and the duration of the operation to be included in the submittal.

14. Heat emission calculations to be submitted to the consultant for approval, installation of central battery system should be in well-ventilated and cooled rooms based on the heat emission calculations.
15. Communication compatibility with BMS and all interfaced systems.
16. Submission of sample board.

Workshop Drawings for Emergency lighting (battery) system.

The consultant should instruct the contractor to submit the workshop drawings for the consultant's approval and then for the authority's approval, including the followings:

1. All Civil Defense comments on design drawings should be replied to and fulfilled.
2. The power supply for emergency lighting should be fed from separate miniature breakers MCB and ELCB (reference circuit no. of power supply should be marked on each emergency lighting circuit).
3. Locations of the system's Control Panels.
4. Network and wiring between control panels.
5. EM lighting fixtures in each loop or zone should be studied in each circuit for loading and voltage drop.
6. EM lighting fixtures and exit lighting should be coordinated with architecture layouts and internal furniture.
7. Emergency lighting grouping, zones and fittings should be coordinated with the architectural layout of the building.
8. Power supply for control panels and central battery system shall be from the emergency supply (DB or SMDB) (if the generator is available).
9. Mounting height of all devices and accessories should be clear on workshop drawings or should be according to the approved schedule of heights.
10. For recessed fittings in false ceilings, installation should be coordinated with false ceiling work and other services.
11. Workshop drawing for emergency lighting system clarifying the location of all circuits connected to CBS (size, type of wire, location of EM lights, type of fitting if surface or recess) to be submitted for consultant's approval, ensuring coordination with all other services. The Workshop drawings should be signed by the suppliers of the material. The signature gives an indication that the supplier specialist is accepting the installation and that the installation details match the supplied system.
12. The Workshop drawings of the riser diagram for emergency lighting system should include all details of the submitted system loops or zones, circuit numbering for power supply, CBU, Emergency lights,

Exit lights, wires' sizing based on voltage drop calculations, connection to BMS and fire alarm system, voltage monitoring relay connection etc.

13. Control room and security room drawings with complete details and dimensions showing the location of the Main Control Panel, monitoring panel and connection to BMS (if any).
14. The contractor should submit coordination drawings based on approved shop showings showing interfacing with other systems and the location of the Emergency, exit and signage lights coordinated with other services (AC grills, normal light, sprinklers, FA detectors, etc..).

Emergency lighting system's workshop drawing should be submitted to Authority (Civil Defense) once it is approved by the consultant. Submission Fees shall be paid to Civil Defense by (Contractor/Client) as specified in project scope.

Installations of Emergency lighting (battery) system and emergency lights.

The consultant should follow the contractor to check the following activities during the installation and fixing of devices for the emergency lighting system. The contractor should be notified of the followings:

1. The supplier must submit the installation method statement to the consultant for his approval, including each piece of equipment, outlet, panel, and light fixture.
2. Activity chart, work progress and look ahead program of work for Emergency lighting system clarifying work execution (civil works, electrical interfacing with other systems, finishing work, testing/ commissioning and defining scope of work with monthly updated).
3. The single-line diagram and workshop drawings should be approved by the consultant and the Civil Defense authority.
4. Containment system, conduits and wiring between all lighting fixture point, main control panel, CBU ... etc. should be installed as per approved workshop drawings.
5. Central Battery System wiring should be installed in trunking, GI conduits, or to be fixed on the ceiling by fire rated strips from the same manufacturer.
6. All devices, wires, junction boxes... etc., should be protected properly during installation.
7. Final location of all Emergency and Exit lights should be located according to the approved workshop drawing and coordination drawings.
8. System components and circuit numbering Identification (devices, wiring...etc.) should be provided by using engraved labels and ferrules.
9. Emergency lighting cable should comply with the project's specification and Civil Defense regulation (Authority), taking into consideration the maximum voltage drop within the permissible limit.

10. Cover to be fixed for every emergency lighting point for protection from dirt due to daily construction work.
11. Control panel should be interfaced with Fire Alarm System and Building management system (BMS) (if applicable).
12. The power supply for the Emergency lighting system control panel should be protected by a fuse or breaker. There should be an Indication light in the control panel to indicate power supply availability. Control panel should be fed from separate Section/ELCB/MCB of emergency DB (if available).
13. When circuit from EM light in one area like staircase is fed from two different sources of normal power supply, the control or switching shall be through voltage monitoring relay.
14. Mounting height and location of EM/EXIT Light fittings should be as per the approved coordination drawing or approved schedule of heights for all Service Areas, Staircase, Basement, Parking areas and reflected false ceiling layout.
15. In case the system is self-contained battery type, the power supply for EM light (circuit) should be directly from Distribution Board without any switch.
16. Ensure that all terminations are robustly done on the correct terminal for all devices, control panels, etc....
17. Installation of the earthing system shall be as per the approved design, supplier specialist or manufacturer recommendations.
18. The site Engineer will continuously monitor the installation activity. Periodical reports should include progress and lookahead and should be submitted to the consultant and the client.
19. All control wires and power cables shall be with extra length for connection to the terminal equipment.
20. Ensure that all Central battery units should be installed in air-conditioned areas. Battery rooms also should be air conditioned and ventilated. (Consult the manufacturer for the required cooling).
21. The contractor should use appropriate back boxes for the system components.
22. Supplier, Contractor's Site Engineer should daily and continuously monitor the installation activities till completion. Periodical progress report showing site status, requirements ...etc. shall be submitted by the contractor.
23. Breakdown for all items related to the system shall be proposed by the contractor for billing purposes; these broken-down percentages should be agreed upon by all parties.

Civil Defense Requirement for Handing Over Emergency Lighting System

A. Approved workshop drawings by the authorities and the consultant.

B. Overview and method of operation (sequence of operation).

C. Brief description of the system's performance.

D. Evacuation procedure.

E. A letter from the supplier (certificate) stating that all work related to the system is completed.

F. Letter to Civil Defense requesting for inspection.

G. Complete identification is required for all rooms, emergency/exit lights, and control panels.

H. Fees for Civil Defense to be paid by the client or the contractor (to follow the project specifications).

I. Undertaking letter (if required).

Operation and Maintenance (O&M) manual content for Emergency Lighting System

The consultant should follow the contractor to submit the O&M manuals for the central battery system, including the following requirements:

1. Description of the project system.
2. How to use the manual.
3. Troubleshooting guide (for emergency breakdown).
4. Safety instruction at work.
5. Start-up and shutdown procedure.
6. An original catalogue highlighting the material used.
7. As built drawing including schematic riser diagram/block diagram.
8. List of spare parts.
9. Procedure for routine maintenance for monthly/1 year/5 years.
10. Authority No Objection Certificate (NOC).
11. Warranty certificate for 2 years (as per project specifications).
12. Testing and commissioning certificates/reports.
13. Emergency information.
14. Manufacturer/supplier information and contact details.
15. Overview and method of operation (sequence of operation).

16. Training for the client team (This could be done with separate training manual submittal based on the project specifications).
17. All guidelines for using all devices/controllers, etc....
18. System programming, settings, and configurations.

Testing and commissioning for Emergency Lighting System.

A. The consultant should instruct the Contractor to complete all pending or delayed work related to the system.

B. Testing and commissioning Method Statements to be submitted by the contractor for the consultant's approval including standard readings and the successful test criteria.

C. Complete the wiring megger test before starting any termination for emergency lighting or Central battery unit.

D. A confirmation letter from the supplier is needed that the emergency lighting system installation is completed and ready for operation and testing.

E. Ensure the followings were done or installed properly by visual checking and inspection:

   1. Normal and emergency wiring were installed and completed in all fittings to respective modules. Power cables were tightened Properly.
   2. The panels are in good condition and without any damage.
   3. All panels and loops are identified and labelled by using engraved labels and ferrules.
   4. Ensure that all components are installed as per approved material submittal and workshop drawings.
   5. Mechanical checking for all components inside the control panel.
   6. The power supply for the main panel is installed and connected properly.
   7. Addressing lighting fixtures is done correctly without duplications.
   8. Wiring for all normal and emergency light fittings is completed in each emergency luminaire and attached to its respective module.

F. Function test

   1. Batteries should be fully charged at least 24 hours prior to starting the testing of the system. Voltage across the batteries terminal to be checked physically using voltmeter ensuring it is within the specified voltages.
   2. Status and function of all modules, batteries, LED, Display, Indicators, Fuses and Breaker shall

be checked when the system is operational.

3. Check the system has been programmed for the type of operation and the number of lights as per design in each circuit/loop or floor. Also, check the control panel and monitoring system.
4. Check all indication lamps are working (on/off) based on system status on/off /fault.
5. All emergency lights are to be programmed as non-maintained except the exit lights to be maintained.

G. Final approval for the system is subject to the following:

1. Successful testing and commissioning of the system.
2. Approval of Civil Defense.

Access Control System (Electrical Doors)

Approval of material submittal for Access control system

The consultant should follow the contractor to arrange periodical meetings between the Electrical engineer and the suppliers of access control systems and all other related systems' suppliers like fire alarm, intercom, BMS, doors' ironmongery and door lock suppliers to clarify the mode of operation during Emergency, fire, or power supply failure, and to discuss interlock and interfaces with other systems, material submittal and work progress. The contractor should be instructed to provide the followings for the material submittal approval:

1. The selected brand or manufacturer is listed in the consultant's approved manufacturer list.
2. Submission of a brief description of the access control system and sequence of operation.
3. Compliance statement to be submitted. The supplier should clearly show his compliance to the specifications, non-complied sections to be also highlighted with justifications.
4. An original catalogue highlighting the submitted components and highlighting the technical points complying with the technical specifications.
5. Block diagram in details for access control system clarifying all connection with Fire Alarm system, BMS, Intercom system, EXIT push button, control panel, hand set inside flat card reader…etc.
6. Valid supplier trade license with expiry date, and ministry of trade certification as a legal supplier of the material in the submittal.
7. Ensure compatibility between the system communication protocol with all other related systems (especially Ironmongery, BMS, FA, CCTV, and all other interfaced systems) and equipment, all suppliers involved in supplying the related systems should sit in a meeting to discuss the communication protocol and to confirm smooth communication.
8. Mode of operation to be included in the submittal with reference to the schematic or block diagram.
9. Flow chart for all related activities of the system and the required interlocks according to the specifications.
10. The submittal should show earthing requirements for the system.
11. Type of sensors or controllers, number of access cards to comply to the specifications and client requirements. Ensure unique access card for all access control systems in the project like Access and Parking barrier, etc. with multi access levels.
12. Coordinate with Architect and door supplier to match the magnetic lock colour and size with the door for colour and installation. If Ironmongery is electrically controlled, coordination with door supplier for provision of the requirements for electrical supply connection by meeting with both suppler (the Access control, and the door supplier).

13. Control panel should be capable for interfacing with Fire Alarm, BMS, intercom systems, doors ironmongery and all other interfaced systems and communicated systems through central networking (if applicable).
14. Details of wiring diagram for control panel including terminal block showing all incoming and outgoing wires from the control panel.
15. Moving sensors to be provided according to the project specifications, sensors should cover all areas without dead zones. The supplier specialist should sign the shop drawings as a reference for correctness.
16. Troubleshooting procedures for emergencies should be submitted in the submittal.
17. Explanation of card programming and the required tools and equipment to be part of the submittal. The programming machine should be part of the system.
18. Magnetic lock and ironmongery colour, specifications, and installation should be coordinated with the architect's work and door supplier.

Workshop Drawings for Access Control System

The consultant should instruct the contractor to submit the workshop drawings for the consultant's approval, including the followings:

1. Preparation of workshop drawing for access control system. A separate detail drawing (plan and elevation) including conduit sizing details, power outlets (fused cable outlet with indicator lamp), control box, intercom system, machine above the door, card reader, push button, Junction box (JB) for intercom system, Connection to BMS, central access moving sensor details and mounting height for all devices.
2. Control room and Security room coordination drawings in detail showing the location of the Main Control Panel, monitoring panel and connection to BMS (if applicable).
3. Discussion with the supplier for the best location of override switches; these locations should be approved by the consultant and maybe the security system or client. Location to be shown int the shop drawings.
4. The final shop drawings should be signed by the material suppliers. The signature gives an indication that the supplier specialist is accepting the installation and that the installation details match the supplied system.
5. Segregation between control data cabling and power wiring should be maintained for safety and noise immunity.

Installation of Access Control System

The consultant should follow the contractor in the installation stage of the Access control System to check and ensure the followings:

1. The installation method statement must be submitted for the consultant's approval by the contractor.
2. Ensure that all conduits and wires are installed between the sliding door, machine above the door, control panel, push button, control room (control panel and BMS), magnetic lock, power outlet and all other interfaces...etc.
3. Proper protection for all wiring, devices, junction boxes...etc. Shall be provided during installation.
4. The final location of all accessories should be as per approved workshop drawings and coordination drawings.
5. Identification for the system (devices, wiring, junction box, control panel ... etc.) by using engraved labels and ferrules shall be provided.
6. Break down for all items related to the system shall be proposed by contractor for billing purpose; these broken-down percentages should be agreed by all parties.
7. Activity chart, work progress and look ahead program of work for Access control system clarifying work execution (civil works, electrical interfacing with other systems, finishing work, testing/commissioning and define scope of work with monthly updated).
8. The power supply outlet for the access control system should be fused, and the indicator lamp outlet on the control panel should indicate power availability. This power supply should be supplied from a separate section (ELCB, MCB) fed from an emergency DB if available.
9. The earthing system shall be connected according to the supplier/manufacturer's recommendation.
10. A wiring diagram for the control panel shall be provided and kept inside the control panel.
11. Moving sensors to be installed according to the project specifications, sensors should cover all areas without dead zones. The supplier specialist should sign the shop drawings as a reference for correctness.
12. Power and control wiring should be selected according to the standard, specification, and authority regulations. Voltage drop calculation should ensure the voltage drop is withing the accepted level.
13. A backup power supply (battery) shall be provided inside the control panel.
14. Installation or fixing of any accessories above the false ceiling is not allowed.
15. Ensure proper and correct termination is done to the correct terminal for all devices, control panel... etc. Cables to be connected correctly and to be checked to the correct torque. Cables to be insulation tested before connection.

16. The mounting height of devices should be as per approved workshop drawings or approved mounting height schedule.

17. Supplier and Contractor's Site Engineer should monitor daily and continuously the installation activity till completion. Periodical progress report showing site status, requirements ...etc. shall be submitted by the contractor.

18. All control wires and power cables shall be with extra length for connection to the terminal equipment.

19. The contractor should use appropriate back boxes for system components.

Testing and commissioning (T&C) for Access Control System.

A. Before starting the testing of the access control system, the consultant should instruct the Contractor to complete all work related to the system (card reader/control panel/push button/sensor…etc.) The contractor should submit a specific method statement for testing and commissioning, including a checklist showing the standard or successful readings and status in the submitted format.

B. Megger test for all circuits (wiring) shall be completed before starting any termination for the system.

C. Confirmation letter from the supplier that the access control system installation is completed and is ready for operation and testing.

D. Visual inspection and check for the system confirming proper fixing, installation and termination are done for the following:

    1. All components are installed as per the approved submittal and workshop drawing.
    2. The main control panel and all accessories (bush button, card reader, moving sensor, door lock…etc.) are installed correctly.
    3. All required power supplies are connected as per system requirements.
    4. Main control panel, push button, …etc. are identified and labeled by using engraved labels and ferrules.
    5. Mechanical checking for all components inside the control panel

E. A function test should be carried out for the system to confirm that the system is working as per the project specifications.

    1) Confirming when the power supply is off, the lock will not open due to the backup power supply battery.
    2) Confirming in case of fire, the lock will open to allow evacuation.
    3) Test the programming of the access cards.

4) Check the card reader to open doors from outside. Use the push button or sensor to test the doors that are opening from inside.

F. Final approval is subject to the successful completion of testing and commissioning of the system.

Operation and Maintenance manuals (O&M) for the Access control system.

The operation and maintenance manuals should include the following:

1. Description of the project system.
2. How to use the manual.
3. Troubleshooting guide.
4. Safety instruction at work.
5. Switching "on" and "off" procedures for the system.
6. An original catalogue highlighting the used material, its part number, and the major technical points according to the specifications.
7. As-built drawing, including wiring diagram for the main control panel and block diagram with all interfacing systems and equipment.
8. Spare parts and special tools List
9. Procedure for routine and preventive maintenance for monthly/1 year/5 years.
10. System and card programming and configurations with software details.
11. Warranty certificate for 2 years (as per project requirement).
12. Testing and commissioning certificates/reports.
13. Emergency procedure and information.
14. Manufacturer/supplier information and contact details.
15. Overview and mode of operation (sequence of operation).
16. Training for the security, operation, and maintenance team. (This can be done by submission of separate training manual including training details and session content).
17. All guidelines for using all devices/controllers… etc.
18. Description of all interfaced and interlocked systems to the Access control system.

Intercom (Interphone) System

Approval of material submittal for Intercom System

The consultant should follow the contractor to arrange periodical meetings between the Electrical engineer and the suppliers of the intercom system and all other interfaced and interlocked systems 'suppliers like access control, BMS, CCTV, door lock and ironmongeries to discuss and clarify the mode of operation, material submittal requirements, work progress ..etc. The contractor should provide the followings for the consultant's approval for the material submittal:

1. The selected brand or manufacturer is listed in the consultant's approved manufacturer list.
2. A brief description of the intercom system and sequence of operation to be submitted for approval.
3. Compliance statement to be submitted. The supplier should clearly show his compliance to the specifications, non-compliant sections should also be highlighted with justifications.
4. The submittal should include the original catalogue highlighting the selected components and the technical points that comply with the project specifications.
5. Valid supplier trade license with expiry date and Ministry of Trade certification as a legal supplier of the material in the submittal.
6. The control panel should be able to interface with all related systems according to the project specifications.
7. Submission of sample board if applicable.
8. Block diagram/ riser diagram in detail for intercom system clarifying all connections with BMS/access control, EXIT push button, control panel, handset inside flat, card reader…etc. (according to project specifications).
9. The mode of operation should be included in the submittal.
10. Identification and engraving labels to be included in the submittal.
11. All Used cables should be shown in the submittal with their catalogues highlighting the used sizes.
12. System earthing based on the manufacturer recommendations to be shown in the submittal.
13. Coordinate with wiring accessories samples. Check and ensure that the doorbell sound is different from the intercom sound in the flat.
14. Wiring diagram for control panel including terminal block showing details for all ingoing and outgoing wires to the control panel.
15. In the case of two entrances, the system should be operated centrally.
16. In case of power supply interruption or activation of the fire alarm system, all emergency doors connected

to the intercom/central access control system should open.

17. The operating power of the system shall be through a transformer or power supply unit (low voltage 24 volts).
18. Communication compatibility with BMS and all interfaced systems.

Preparing Workshop drawings for the Intercom System

The consultant should instruct the contractor to submit the workshop drawings for the consultant's approval, including the following:

1. Workshop drawing for intercom and access control system (plan and elevation) for main entrance and electrical room clarifying the size of conduit and wires, power supply outlet with indicator lamp, control box, speakers or handset installations and mounting height of all devices... etc.
2. Shop drawings are to be signed by the intercom supplier specialist and all other interfaced systems suppliers. The signature confirms that the supplier specialist is accepting the installation and that the installation details match the supplied system.
3. Coordination drawing in detail for the location of the main panel installation showing installation details coordinated with other services, Doors, Emergency doors, furniture, and architectural work.
4. Power supply for intercom system should be from fused outlet with indicator lamp on control panel indicating availability of power supply and from separate ELCB/MCB and from emergency DB if available.
5. Ensure that the Guardroom, reception and control room have intercom points and override switches (follow the project specifications).
6. Earthing of the system shall be according to the supplier's or manufacturer specialists' recommendation.
7. Single line diagram should indicate the location of handset, master station, control panel, Power supply reference interface with access control system and push button ... etc. with reference to the floor level in the building.
8. There shall be at least one point for intercom inside the guard room and all service rooms like main Electrical room, mechanical rooms ... etc.
9. In the case of two entrances, the system should be operated centrally.

Installation of Intercom System and drawings

The consultant should follow and instruct the contractor in the installation stage of the intercom and its interfaced systems to check and ensure the followings:

1. The supplier should submit an installation method statement for the whole system with all related interfaces and components.
2. Ensure installing all conduits and wires between the electrical door, push button, control panel of intercom system and handset ... etc.
3. Ensure the contractor is using appropriate back boxes for the used components.
4. Proper protection for all devices, wires, and junction boxes shall be provided during installation.
5. Final location of all handsets, push button, main panel ... etc. should be as per approved workshop and coordination drawings.
6. Identification and numbering for all handset, wiring, junction box and other devices shall be done by using engraved labels and ferrules.
7. Activity chart, work progress and look ahead program of work for Intercom System clarifying steps for executing the work including Civil Work, Electrical Work, Finishing Work, testing and commissioning and defining the scope of work for the involved contractors with monthly updates.
8. All control wires and power cables shall be installed as per specifications and standards, voltage drop calculation shall be considered.
9. Break down for all items related to the system shall be proposed by contractor for billing purpose; these broken-down percentages should be agreed by all the parties.
10. Ensure installation of the Earthing for the system according to the supplier's or manufacturer specialists' recommendation.
11. A wiring diagram shall be provided inside the control panel.
12. The width of the junction box behind the handset should be less or equal to the width of the handset.
13. The height of all handsets, push button, main unit ... etc. shall be as per approved workshop and coordination drawings.
14. The opening code list is to be forwarded to the client in handing over time.
15. Ensure that all termination is robustly done on the correct terminal for all devices, control panel... etc.
16. Supplier, Contractor's Site Engineer should daily and continuously monitor the installation activity till completion. Periodical progress report showing site status, requirements ...etc. Shall be submitted by the contractor.
17. Segregation between control data cabling and power wiring should be maintained for safety and noise immunity.
18. All control wires and cables shall be with extra length for connection to the terminal equipment.

Testing and commissioning of Intercom System.

A. The consultant should instruct the Contractor to complete all pending or delayed work related to the system (handset/control panel/push button …etc.). Testing and commissioning Method Statements to be submitted by the contractor for the consultant's approval including standard readings and the successful test criteria.

B. Complete the wiring megger test before starting any termination for the system components.

C. A confirmation letter from the supplier is needed confirming that the Intercom System installation is completed and ready for operation and testing.

D. Visual inspection and check for the system confirming that proper installation and termination are done for the followings:

1. All components are installed as per approved submittal, workshop, and coordination drawing.
2. The installation of the main entrance panel, handsets inside flats and reception, junction box in riser for all floors and inter-phase with access control system are completed.
3. The power supply has been connected according to the system requirement.
4. Main Control Panel, handset, junction box …etc. are identified and labeled by using engraved labels and ferrules.

E. Function test to ensure the system is working according to the project specification:

1. Operate the system to check flats dial numbers, sound level in the flat entrance panel and operation of switch board in the reception/control room.
2. Check the doorbell sound is different than the intercom call sound.

F. Final approval for the system is subject to successful testing and commissioning of the system.

Operation and Maintenance (O&M) manual content for Intercom System

The consultant should follow the contractor to submit the O&M manuals for the Intercom system, including the following requirements:

1. Description of the project system.
2. How to use the manual.
3. Description of the interfacing system to access control doors and any other interfaced systems (electric door/sliding door).
4. Troubleshooting guide for failures and breakdowns.
5. Safety instruction at work.

6. Start-up and shutdown procedure.
7. An original catalogue highlighting the technical specifications and the material used with part no.
8. The as built drawing includes a block diagram and a wiring diagram for the main control panel.
9. List of spare parts and special tools.
10. Procedure for routine and preventive maintenance for monthly/1 year/5 years.
11. System programming, configurations, and software (if applicable).
12. Warranty certificate for 2 years (as per project requirement).
13. Testing and commissioning certificates/reports.
14. Emergency procedure information.
15. Manufacturer/supplier information and contact details.
16. Overview and method of operation (sequence of operation).
17. Training for the security, operation, and maintenance team. (Based on the project, it may be required to submit a training manual for approval of training content).
18. All guidelines for using all devices/controllers… etc.

CCTV System

Material Approval for closed circuit Television (CCTV)

The consultant should follow the contractor in arranging periodical meetings between the electrical engineer and the suppliers of CCTV and other interfaced systems to the CCTV Ex. Access control system, BMS, Gate Barrier, fire alarm... etc., to discuss and clarify the mode of operation, material submittal requirements, work progress, recording method and the latest Authority (Police, security) requirements (if there are any)..etc. The contractor should provide the followings for the consultant's approval for the material submittal:

1. The selected brand or manufacturer is listed in the consultant's approved manufacturer list.
2. A brief description of the CCTV system and sequence of operation.
3. The submittal should include the original catalogue highlighting the selected components and the technical points that comply with the project specifications.
4. Compliance statement to be submitted. The supplier should clearly show his compliance to the specifications, non-complied sections to be also highlighted with justifications.
5. Submission of sample board.
6. Valid supplier trade license with expiry date, and ministry of trade certification as a legal supplier of the material in the submittal.
7. The martial submittal for the CCTV system should include a system description and sequence of operation.
8. System earthing should be included based on the manufacturer's requirements.
9. Equipment and component identifications with engraved labelling should be included in the submittal.
10. Control wiring to be included in the submittal.
11. Flow chart for all cameras, including the location, type, mounting type, power supply reference ... etc. and any other interfaced system with CCTV, according to the project specification and Authority (Police or security system) requirements.
12. Block diagram for CCTV system showing all system details including all activities connected to the system like (power supply reference, connection to BMS, connection to the police office, connection to gate barrier, connections to the monitors, DVR computers and interface with all other systems ...etc.)
13. Camera types to be selected based on the specifications, installation location (outdoor and indoor) and type of installation (pendant, ceiling mounting, outdoor mounting with bracket Etc.).
14. Wiring diagram for the Control Panel, DVR... etc. including terminal block showing details for all ingoing and outgoing wires to the control panel.

15. If motion detector for the system is required. Motion detector to be installed based on supplier specialist guidance and component specifications and catalogue. (Motion detector technical sheet to be included in the submittal).
16. All cameras shall have built-in transformers according to the project specifications. Separate transformer will create installation and maintenance difficulties as it should be hidden.
17. Communication compatibility with BMS and all interfaced systems.

Workshop drawings for CCTV System

The consultant should instruct the contractor to submit the workshop drawings for the CCTV system for the consultants and related Authority approvals, including the followings:

1. Workshop drawings for CCTV system should be submitted with power drawing including all installation details (plan and sections). The drawings should show the interfaced systems, containment and components which are connected to CCTV system like BMS, access door, gate barrier, power supply, Fire Alarm and lifts etc. showing the mounting height of all devices, camera…etc. with complete coordination with all other MEP services.
2. Coordination drawing and details for the Control room and security room showing the location of main equipment of CCTV with furniture layout including location for main control panel, digital video recorder DVR, monitors… etc., with all connections to the other interfaced systems, like BMS, fire alarm … etc.
3. Shop drawings to be signed by the suppliers of the material. The signature gives indication that the supplier specialist is accepting the installation and the installation details is matching the supplied system.
4. Power supply for CCTV system (Cameras and Control Panel) should be as per supplier requirements and to be from separate section (ELCB and MCB) in the dedicated final distribution board, using fuse connected outlet. The panel should have indicator light for power availability. Power supply should be provided from emergency DB if available.
5. Coordination with normal lighting: Consult the supplier specailist to study the lighting level in the areas of camera installations and to confirm that lighting/brightness around the cameras is good and enough. (Site visit is required or lighting drawings to be studied).

Installation of CCTV system

The consultant should follow and instruct the contractor in the installation stage of CCTV and its interfaced systems' contractors to check and ensure the following: -

1. Supplier should submit installation method statement for the whole system with all related interfaces and components for consultant's approval.
2. Ensure installing all conduits and wires between all cameras, Main Control Panel monitors, and all other components related and connected to interfaced systems.
3. All control wires and power cables shall be installed correctly according to the approved shop and coordination drawings with extra length for easy connection to the equipment terminal.
4. Proper protection for all devices, wires, junction boxes…etc. It should be provided during installation work.
5. The mounting height of all cameras should be as per approved workshop drawings or approved mounting height schedule.
6. The final location of the camera and power supply should be according to the approved coordination drawings and shall be coordinated with all other services and furniture layout, like (AC duct, FCU, light fittings, cable tray, drop beam and location of furniture…. Etc.)
7. The control panel should be capable of interfacing with all related systems based on project specifications, authority, and client requirements.
8. All control wires and power cables shall be installed as per specifications and standards, voltage drop calculation shall be considered.
9. CCTV earthing installation should follow supplier specialist and manufacturer recommendations.
10. Camera identification and numbering, wiring, junction boxes, and other devices shall be done using engraved labels and ferrules.
11. All cameras and power supplies should be installed in accessible areas for easy maintenance.
12. All cameras in the parking area or outside the building should be installed far away from the HV cable.
13. Ensure that all termination is robustly done on the correct terminals for all devices, cameras, control panels, etc.
14. Supplier, Contractor's Site Engineer should continuously monitor the installation activity. Periodical reports should include progress and lookahead and should be submitted to the consultant and the client.
15. Break down for all items related to the system shall be proposed by contractor for billing purpose; these broken-down percentages should be agreed by all parties.
16. Activity chart, work progress and look ahead program of work for CCTV system clarifying steps for executing the work (civil, electrical, finishing work, testing, and commissioning) and defining scope of work for the involved contractors with monthly updates.

17. Segregation between control, data cabling and power wiring should be maintained for safety and noise immunity.

18. Ensure all installed materials are according to the approved material submittal and are free from damages due to storing or transportation.

19. Ensure that the contractor is using suitable back boxes for all components of the system.

Requirements related to external work.

20. Routing of power and control wires should be coordinated with external services and should avoid HT, LT, BMS and telephone cables Etc.

21. The final location of outdoor cameras should be decided after consulting the supplier specialist.

22. Coordinate with electrical work for power supply source, voltage drop, and signal attenuation.

23. Coordinate with an irrigation system to avoid water splash on cameras.

24. Install earthing conductors according to manufacturer requirements.

25. If the camera to be fixed with outside light fitting, extra sleeves to be provided for camera power and signal in addition to the lighting sleeves.

Testing and commissioning for CCTV system

A. The consultant should instruct the Contractor to complete all pending or delayed work related to the system Testing and commissioning. T&C Method Statements should be submitted by the contractor for the consultant's approval, including standard readings and successful test criteria.

B. Complete the wiring megger test before starting any termination for the system components.

C. A confirmation letter from the supplier is needed confirming that the CCTV system installation is completed and ready for operation and testing.

D. Visual inspection and check for the system confirming that proper installation and termination are completed for the followings:

   1) Main equipment inside the control room (monitors, DVR, VCR, Multiplexer … etc.) and all cameras.

   2) Location of all cameras and types as per approved workshop drawings.

   3) The power supply has been connected according to system requirements.

   4) Main control panel, monitors, DVR …etc. are identified and labeled by using engraved labels and ferrules.

   5) Correct monitor installation in the control room at the specified height according to the approved shop and coordination drawings.

E. Function Test

1) Ensure that all cameras are displayed on the monitor.

2) Ensure that by using the multiplexer, matrix switcher, operation, and camera selection are functioning.

3) Ensure that the recorder and playback quantity are ok.

4) Ensure that the identification of all cameras is displayed on the monitor.

5) Ensure that all cameras (fixed/PTZ) cover the required area.

6) Ensure that all interfaced systems are reacting correctly.

F. Final approval for the system is subject to:

1) Successfully testing and commissioning.

2) Authority Approval.

Operation and Maintenance Manual for CCTV systems

The consultant should follow the contractor in submitting the O&M manuals for the CCTV system, including the following requirements.

1. Description of the project system.
2. How to use the manual.
3. Troubleshooting guide.
4. Safety instruction at work.
5. Start-up and shutdown procedure.
6. An original catalogue highlighting the used material with part number and highlighting the technical points and specifications.
7. As built drawing including wiring and block diagram.
8. List of spare parts.
9. Procedure for routine and preventive maintenance for monthly/1 year/5 years, etc.
10. How to make programming, settings for the system and system configurations
11. Warranty certificate for 2 years (as per project requirement).
12. Testing and commissioning certificates/reports.
13. Emergency information.
14. Manufacturer/supplier information and contact details.

15. Overview and method of operation (sequence of operation).
16. Training for security, operation, and maintenance teams. (Training can be done by submitting a training manual for the consultant and client approval).
17. All guidelines for using all devices/controllers, etc....
18. Description of all interfacing systems and the communication method.

Traffic Barrier System

Approval of the material submittal for the Traffic Barrier System.

The consultant should follow the contractor to arrange periodical meetings between the Electrical engineer and the supplier of the traffic barrier system and all other interfaced and interlocked systems suppliers like fire alarm, BMS, intercom system and CCTV system to discuss and clarify the mode of operation, material submittal requirements, work progress and the latest Authority requirement ..etc. The contractor should provide the followings for the consultant's approval for the material submittal:

1. The selected brand or manufacturer is listed in the consultant's approved manufacturer list.
2. A brief description of the traffic barrier system and sequence of operation.
3. The submittal should include the original catalogue highlighting the selected components and the technical points that comply with the project specifications.
4. Compliance statement to be submitted. The supplier should clearly show his compliance to the specifications, non-complied sections to be also highlighted with justifications.
5. Valid supplier trade license with expiry date, and ministry of trade certification as a legal supplier of the material in the submittal.
6. The mode and philosophy of operation should be submitted with the submittal.
7. Block diagram for traffic barrier system shall include details for all interfaced or interlocked activities connected/related to the system like power supply reference, connection to BMS, CCTV, FAS, intercom, manual push button, magnetic contact ... etc.

1. Wiring diagram for control panel including a terminal block diagram showing details for all ingoing and outgoing wires to the control panel.
2. Communication compatibility with BMS and all interfaced systems.
3. System earthing is to be submitted based on the manufacturer's requirements.
4. Submission of the controller and entry cards according to the project specifications. Coordination with Access control system for using the same card with the Traffic barrier System. (If Applicable).
5. The gate barrier aluminium arm should be phosphorescent red strips supported from the end of the arm.
6. The system should include a safety photocell.
7. The card programming machine should be part of the system (built-in). (To be coordinated with the Access control system).

8. The control panel should be capable of interfacing with the Fire Alarm System, BMS, security intercom CCTV and all related systems according to the project specifications.
9. All control wires and power cables shall be installed as per specifications and standards, voltage drop calculation shall be considered.

Workshop Drawing for Traffic Barrier System

The consultant should instruct the contractor to submit the workshop drawings for the Traffic Barrier System for the consultant's approval, including the following:

1. Workshop and coordination drawings for the traffic barrier system should clarify and indicate the location of the isolator, intercom system, foundation, length of the barrier, safety photocell, magnetic contact, routing of all conduits between all components, power supply reference, and interface with all other related systems in the project like (BMS, CCTV, FAS, Intercom, Access control etc.). The shop drawings should include sketches with dimensions that can allow the longest car to enter the building.
2. Control room/Security room coordination drawing and details showing the location of all components related to the system, like override switch (push button) monitoring panel, interphone system and connection to BMS (if any) with clear dimensions.
3. Workshop drawings to be signed by the supplier of the material. The signature gives indication that the supplier specialist is accepting the installation drawings and the installation details are matching the supplied system requirement.
4. Power supply for traffic barrier system should be from separate Section/ELCB/MCB and from emergency DB if available.
5. Traffic Barrier earthing connections should comply with the supplier's/manufacturer's recommendation.
6. Override switch (in the control room, reception, security room guard room, etc.) to be provided based on the supplier specialist recommendations and to be submitted for consultant and client approval.

Installation of Traffic Barrier System

The consultant should follow and instruct the contractor in the installation stage of the Traffic Barrier System and its interfaced systems to check and ensure the followings:

1. The supplier should submit an Installation method statement for the whole system with all related interfaces and components.
2. All conduits and wires shall be installed between the traffic barrier, security intercom, FAS, BMS, power supply (isolator and DB), push button in the control room, CCTV … etc.
2. Proper protection for all wiring and devices shall be provided during installation.

3. The final location of all accessories should be as per the approved workshop and coordination drawings.
4. Identification and numbering for system components (isolator, wiring, junction box, terminal block for control panel… etc.) shall be provided using engraved labels and ferrules.
5. Before installing the equipment, ensure all equipment are free of damages due to storing and transportation, and ensure all cables and wiring are healthy (do insulation test by megger).
6. The contractor should use appropriate junction and back boxes for all system components.
7. Traffic Barrier earthing connections shall be made as per the supplier's/manufacturer's recommendation.
8. The power supply isolator should be very close to the traffic barrier for emergency and maintenance purposes.
9. Ensure that all termination is robustly done on the correct terminals for all devices, control panel … etc.
10. Activity chart, work progress and look ahead program of work for Traffic Barrier System clarifying steps for executing the work including Civil and road pavement Work, Electrical Work, Finishing Work, testing and commissioning, and defining the scope of work for the involved contractors with monthly updates.
11. Supplier, Contractor's Site Engineer should daily and continuously monitor the installation activity till completion. Periodical progress report showing site status, requirements ...etc. Shall be submitted by the contractor.
12. Segregation between control/ data cabling and power wiring should be maintained for safety and noise immunity.
13. All control wires and power cables shall be with extra length for connection to the terminal equipment.
14. Breakdown for all items related to the system shall be proposed by the contractor for billing purposes; these broken-down percentages should be agreed upon by all parties.
15. Override switches (in the control room, reception, security room guard room Etc.) were installed according to the supplier specialist recommendations and coordinated with all other services.

Testing and commissioning of Traffic Barrier System.

1. Before starting the testing of the traffic barrier system, the consultant should instruct the Contractor to complete all work related to the Traffic Barrier system and all related interfaced systems. The contractor should submit a specific method statement for testing and commissioning, including a checklist showing the standard or successful readings and status in the submitted format.
2. Megger test for all circuits (wiring) shall be done before starting any termination for the system.
3. Confirmation letter from the supplier confirming that the traffic barrier system installation is completed,

and it is ready for operation and testing.

4. Visual inspection and check for the system to confirm that proper fixing, installation, and termination are done for the followings:
    1. Confirming the main control panel, barrier cabinets, push button, card reader, power supply isolator, magnetic contact, inter-phase connection, safety edge, inter-phone system, safety photocell ... etc. all are selected as per approved submittal and installed according to the approved workshop and coordination drawings.
    2. The connection of the power supply and earthing system is as per the standard, manufacturer recommendations, and authority requirements.
    3. The main control panel is dressed neatly, and the wiring diagram is included in the panel. Termination of incoming and outgoing wires (terminal block) is completed with proper identification (engraved labels and ferrules).
    4. A Smart Card Programming Machine is provided as a part of the control panel.
    5. All loop detector wiring is done in the entrance/ exit and is covered with mastic.
5. Function tests should be carried out on the system after installation and connection are completed. It should include:
    1. Operation of the system in case of power failure.
    2. Operation of the system in case of fire (interface with Fire Alarm System).
    3. Operation of the system in normal mode using card reader and override switches in the control room. (Timer programming of the incoming and outgoing cars as per specification and interfacing with BMS, CCTV, security, and intercom).
    4. Operation of safety loop and photocell.
    5. Card generation, programming and reading.
6. Final approval for the system is subject to successful testing and commissioning of the system.

Operation and Maintenance Manual for Traffic Barrier System.

The operation and maintenance manuals should include the following:

1. Description of the project system.
2. How to use the manual.
3. Troubleshooting guide.
4. Safety instruction at work.

5. Start-up and shutdown procedure.
6. An original catalogue highlighting the material used with part numbers and highlighting the compliance to technical requirements in project specifications.
7. As built drawing including wiring diagram for the control panel and all other components, block diagram and all interfaces.
8. List of spare parts and special tools.
9. Procedure for routine and preventive maintenance for monthly/1 year/5 years.
10. System programming, settings, configurations, card reader programming, and the software used, if applicable.
11. Warranty certificate for 2 years (as per project specifications).
12. Testing and commissioning certificates/reports.
13. Emergency information and manual operation.
14. Manufacturer/supplier information and contact details.
15. Overview and method of operation (sequence of operation).
16. Training for the security, operation, and maintenance team. (This can be done by submission of separate training manual including training details and session content).
17. All guidelines for using all devices/controllers… etc.
18. Description of all interfacing systems like fire alarms., CCTV … etc.

Public Address System

Approval of material submittal for Public Address System

The consultant should follow the contractor in arranging periodical meetings between the electrical engineer and the supplier of the public address system and all other interfaced systems' suppliers, e.g., the Fire Alarm system. The meeting should discuss and clarify the mode of operation during a fire, the latest Civil Defense requirements, interlocks and interfaces with other systems, material submittal and work progress. The contractor should be instructed by the consultant to provide the followings to get the consultant's approval for the material submittal:

1. The selected brand or manufacturer is listed in the consultant's approved manufacturer list.
2. A brief description of the public address system and sequence of operation.
3. Compliance statement to be submitted. The supplier should clearly show his compliance to the specifications, non-complied sections to be highlighted with justifications.
4. The submittal should include the original catalogue highlighting the selected components and the technical points that comply with the project specifications.
5. The martial submittal for the system should include a system description and sequence of operation.
6. Valid supplier trade license with expiry date, and ministry of trade certification as a legal supplier of the material should be included in the submittal.
7. System earthing should be included in the submittal based on the manufacturer's requirements.
8. Flow chart for all public address system devices and components including the type, rating, mounting type, power supply reference circuit, speakers…etc. with all interfaced systems components like BMS, Fire Alarm System …etc. According to the project specifications and authority requirements.
9. Riser diagram for public address system in detail, including all activities connected to the system like (power supply reference, connection to BMS, Fire Alarm System and main equipment in control room… etc.).
10. All indication lamps should be LED.
11. Block diagram for public address system showing all systems details including all activities connected to the system like (power supply reference, connection to BMS, FAS, main control in control room …etc.).
12. The control panel should be capable of interfacing the system with Fire Alarm and BMS systems (and any other system based on project specifications).
13. Communication compatibility with BMS and all interfaced systems.

Workshop Drawings for Public Address System

A. The consultant should instruct the contractor engineer to submit the workshop and coordination drawings for consultant approval, including the followings: -

1. All comments by Civil Defense on the design drawing of the Fire Alarm System, including the Public Address and voice evacuation system, should be considered in the proposed PA system for evacuation during fire.
2. Each circuit (zone/loop) in the public address system should have 10% capacity spare for additional loudspeakers in future.
3. Power supply for public address system should be fed from separate section, MCB / ELCB and from emergency DB if available.
4. The location of a central equipment rack, power amplifier, speaker, sounder, microphones...etc, should be shown in the coordination drawing and coordinated with other services and furniture layout.
5. The communication network between control panels and other components and systems.
6. Number of speakers in each zone/ circuit and loop should be studied and coordinated with the rating/ capacity of the system zones or channels and considering signal attenuation for far speakers. Each circuit should have 10% additions for future extension.
7. Speakers' grouping selection in each circuit should be studied based on the Architectural layout.

B. Workshop and coordination drawings for the public address system should be part of the Fire Alarm shop drawing, clarifying the location of all speaker circuits, and main equipment inside the control room with complete coordination with all other electromechanical services and architectural work.

C. Control room, Security room workshop and coordination drawings with all details showing location of Main Control Panel, rack, monitoring panel and connection to Fire alarm system. (Voice evacuation system).

D. Shop drawings to be signed by the suppliers of the material. The signature give indication that the supplier specialist is accepting the installation drawings and the installation details are matching the supplied system.

E. Works shop drawings should be submitted to the Authority (Civil Defense) for approval with the voice evacuation system after being approved by the consultant. Submission fees should be paid by (Contractor/Client).

F. Activity chart, work progress and look ahead program of work for public address system clarifying steps for executing the work (Defining scope of work, civil work, electrical work, finishing work, testing, and commissioning with monthly updates).

Installation of Public Address System

The consultant should follow the contractor to ensure installing all devices and ensuring the following activities for the public address system with all related and interfaced systems and components:

1. The contractor should submit an installation method statement for the whole system with all related interfaces and components.
2. Before installation, ensure all materials are as approved in the submittal and without any damages.
3. The contractor should use appropriate back boxes for all components in the installed system.
4. Ensure that all conduits and wires are installed between speakers, the main control system, interfaced systems, the admin desk, etc.
5. Proper protection for all devices shall be provided during installation work.
6. Speakers Identification and numbering, wiring, junction boxes and other devices shall be done by using engraved labels and ferrules.
7. Ensure that all terminations are done correctly and robustly on the correct terminals for all devices, control panels and speakers etc....
8. The final location of speakers should be according to the approved workshop and coordination drawings. For speakers in large spaces and volumes, speakers' locations should be distributed by the manufacturer or specialized supplier using suitable software, and the designer should sign the drawing when it is completed.
9. All wiring for control and power to be as per standards, project specification and authority regulation. Wires to be tested before installation.
10. A cover and adapter should be fixed for every speaker point for protection.
11. The mounting height and location of speakers, control panel, and rack shall be according to the approved coordination drawings and approved height schedule.
12. Earthing for the system shall be done according to supplier/manufacturer recommendation.
13. Controlling and Monitoring modules such as tuner, tape player, CD player, Reo matrix, switcher, amplifier shall be installed as per manufacturer's recommendations.
14. Proper ventilation for the rack should be provided to avoid over heating (consult the manufacture or supplier specialist).
15. Wiring diagram shall be provided inside the control panel.
16. Supplier, Contractor's Site Engineer should daily and continuously monitor the installation activity till

completion. Periodical progress report showing site status, requirements ...etc. Shall be submitted by the contractor.

17. Segregation between control, data cabling and power wiring should be maintained for safety and noise immunity.
18. All control wires and power cables shall be with extra length for easy connection to the equipment terminal.
19. Breakdown for all items related to the system shall be proposed by the contractor for billing purposes; these broken-down percentages should be agreed upon by all parties.

Civil Defense Requirements for handing over of Public Address System.

The consultant should follow the contractor to fulfil the latest civil defence requirements and regulations, however, the following requirements should be provided for civil defence approval:

1. Approved workshop drawings by the consultant.
2. Overview and method of operation (sequence of operation).
3. Brief description of the system's performance.
4. A letter from the supplier (certificate) confirming that all installation work for the system has been completed.
5. An official letter was sent to civil defence requesting system inspection.
6. Complete identification for all speakers and control panels.
7. Fees payment.
8. Undertaking letter (if required).

Testing and Commissioning of Publica Address System

A. The consultant should instruct the Contractor to complete all pending or delayed work related to the system. Testing and commissioning method Statements to be submitted by the contractor for the consultant's approval including standard readings and the successful test criteria.

B. Complete the wiring megger test before starting any termination for the system components.

C. Confirmation letter from the supplier stating that the PA System installation is completed and ready for operation and testing.

D. Visual inspection and check for the system to ensure that proper installation and termination are completed and confirming the followings

   1. Main equipment inside control room (amplifier, switcher, player, Rio matrix…etc.) and all speakers are installed according to the approved workshop/ coordination drawing.

2. The locations of all speakers are as per the approved workshop drawing.
3. The power supply is connected as per system requirements.
4. Main control panel, amplifier, switches .. Etc. are identified and labeled by using engraved labels and ferrules.
5. Confirming that the system is procured according to the approved material submittal.
6. System components and interface with fire alarm, BMS, and all other interfaced systems according to the project specifications are installed and completed according to the approved workshop and coordination drawings.

E. Function test to be done ensuring the following:

1. All speakers are working properly.
2. All speakers are set at required sound level and volume controls are adjusted as required without any disturbances or hum.
3. The monitor panel for sound output and source selection is working properly.
4. Tuner, cassette…etc. Are working properly.
5. Operation for each terminated loop as per approved drawings.
6. Microphones in reception and control room music shall stop while paging.

F. Final approval for the system is subject to the following:

1. Successfully testing and commissioning for the system by a third party according to the authority regulations.
2. Approval of Civil Defense.

Operation and Maintenance Manual for Public Address System

1) Description of the project system.
2) How to use the manual.
3) Troubleshooting guide.
4) Safety instructions at work.
5) Start-up and shutdown procedure.
6) An original catalogue highlighting the used material with part number.
7) The as built drawing includes a wiring diagram for the main control panel and a block diagram.
8) List of spare parts and special tools.

9) Procedure for routine or preventive maintenance for monthly/1 year/5 years, etc.
10) Authority NOC.
11) Warranty certificate for 2 years (as per project requirements).
12) Testing and commissioning certificates/reports with successful results.
13) Emergency information.
14) Manufacturer or supplier information and contact details.
15) Overview and method of operation (sequence of operation).
16) Training for security, operation, and maintenance teams. (Training can be done by submitting a training manual for the consultant and client approval).
17) Programming, set up, and system configurations with used software.
18) All guidelines for using all devices/controllers.
19) Description of interfacing with all other systems like FAS, BMS, etc…

MATV System

Material Approval for Master Antenna Television (MATV) System

The consultant should follow the contractor in arranging periodical meetings between the Electrical Engineer and MATV system Supplier to discuss and clarify the mode of operation, system type, number of channels, number of dishes, and any other specified requirements. The contractor should provide the followings for the consultant's approval of material submittal:

1. The selected brand or manufacturer is listed in the consultant's approved manufacturer list.
2. A brief description of the MATV system and sequence of operation.
3. Compliance statement to be submitted. The supplier should clearly show his compliance to the specifications, non-complied sections to be highlighted with justifications.
4. The submittal should include the original catalogue highlighting the selected components and the technical points that comply with the project specifications.
5. Valid supplier trade license with expiry date, and ministry of trade certification as a legal supplier of the material should be shown in the submittal.
6. Riser diagram for MATV system clarifying in detail all Zones, Circuits, Power supply, control panel and all related components ...etc.
7. Supplier confirmation of the possibility of future upgrading of the selected system.
8. List of the required satellites and TV channels in the system based on project specifications and should be approved by the client.
9. The cables and socket outlets should be included in the submittal by including the original catalogues and highlighting the specifications and the selected items.
10. The identification of components and parts with engraved labelling should be part of the submittal.
11. The control panel should be provided with an Indicator light on the control panel indicating the availability of the power supply.
12. System earthing is based on the manufacturer's requirements.
13. Communication compatibility with BMS and all interfaced systems.

Workshop drawings for MATV System.

Workshop drawing should be submitted for approval, including the following: -

1. The location of all dishes/antennas should be shown on roof coordination drawings. The location of the antennas and dishes should be studied carfully regarding signal intruption due to future high rise building installation.

2. Power supply for MATV system should be fed from separate section with breaker and ELCB, reference circuit number of power supply should be marked on the drawing.
3. Location of Main and Sub Control Panel inside control room/electrical room.
4. Network between control panels.
5. Details of an earthing system for the MATV system should be done according to supplier recommendation.
6. Foundation details of the antenna and terrestrial mast location on the roof with dimensions, considering spare foundation for future with complete coordination with all services on the roof.
7. The location of all MATV points should be close to the power supply. These locations should be according to the approved coordination drawings and coordinated with other services and furniture layouts. IP for component should be selected based on installation location.
8. The power supply for the main and sub-control panels should be with a fused switch (connected unit). The power supply should be from separate sections/ELCB/MCB.
9. The mounting height of all devices should be according to the approved workshop drawing or approved mounting height.
10. Flow chart for all control panels clarifying the location, Power supply details, Number of zones, Circuit… etc. For each control panel.
11. Workshop and coordination drawings for MATV system clarifying the location of panels related to the system coordinated with all other services. The location of dishes on the roof should be coordinated with architectural work to hide them from being seen from a far distance.
12. Coordination drawing for control rooms drawing with full details and dimensions showing location of Main and Sub Control Panel with all equipment.
13. Shop drawings should be signed by the material suppliers. The signature gives an indication that the supplier specialist is accepting the approved installation drawings and that the installation details match the supplied system requirements.

Note: In case there is a central system for the compound community, coordination should be done with all other suppliers (irrigation, low current systems, and related services authorities).

1. The contractor should ensure no interference of the electromagnetic field between HT/LT cables and the signal of the MATV system.
2. Care should be taken that no irrigation water splash reaches the equipment of the MATV system (MD and MDU) when equipment panels are external.
3. To consider power supply for MD and MDU.

Installation of MATV System

The consultant should follow up and monitor the contractor during the installation stage of the MATV outlets and control panel for the MATV system. Contractors to check and ensure the followings:

1. The supplier should submit an installation method statement for the consultant's approval.
2. Before starting the installation, ensure all received materials are complied to the approved material submittal.
3. Installation of all conduits and wires between all outlet, junction box for splitter and Control Panel. ...etc. should be done according to approved workshop and coordination drawings.
4. The contractor should use appropriate back boxes for all components in the installed system.
5. Proper protection for all devices, wires, junction boxes... etc., should be provided during installation.
6. The final location of all MATV outlets and junction boxes for the splitter inside flats should be according to the approved workshop and coordination drawings.
7. Identifications for the system (devices, wiring...etc.) shall be made using engraved labels and ferrules.
8. Co-axial cable should be used for IF, RF or any other system should comply to project specification and applicable regulations to ensure signal strength.
9. The mounting height and location of the MATV outlet should be as per the approved coordination drawing or approved mounting height schedule.
10. All accessories, devices and head-end equipment related to the system's components from control panel should comply to the approved workshop and coordination drawing. Connections between components should be done by approved co-axial cable.
11. Ensure correct and robust termination for all devices, control panels, etc.
12. Installation of the earthing system shall be done as per the supplier/manufacturer's recommendation.
13. Segregation between control data cabling and power wiring should be maintained for safety and noise immunity.
14. All control wires and power cables shall be with extra length for easy connection to the equipment terminal.
15. Activity chart, work progress and look ahead program of work for MATV system clarifying steps for executing the work (civil, electrical, finishing work, testing, and commissioning) and defining scope of work for the involved contractors with monthly updates.
16. Supplier and Contractor's Site Engineer should monitor daily and continuously the installation activity till completion. Periodical progress report showing site status, requirements ...etc. Shall be submitted by the contractor.

17. Break down for all item related to the system shall be proposed by contractor for billing purpose; these broken-down percentages should be agreed by all parties.

Testing and commissioning for MATV System

A. Before start testing of MATV system, the consultant should instruct the Contractor to complete all pending or delayed work related to the system. Testing and commissioning method Statement to be submitted by the contractor for the consultant approval including standard readings and successful test criteria.

B. Complete the wiring megger test before starting any termination for the system components.

C. Confirmation letter from the supplier stating that the MATV system installation is completed and ready for operation and testing.

D. Visual inspection and check for the system to ensure that proper installation and termination are completed and to ensure the followings:

   1. All components (Control panel, amplifier, dishes, TV outlets, splitters, etc..) were installed according to the approved submittal, workshop, and coordination drawings.
   2. Proper and correct fixing, installations, and tightness of the co-axial cable, which is terminated properly with LNBs, dishes, head-end amplifiers, splitter and socket outlet. Components are identified using engraved labels and ferrules.
   3. The panel configuration is correct, and the panel is free from damage.
   4. The correct direction of dishes/antennas is using suitable equipment and testers.
   5. All head – end equipment was installed according to approved drawings and submittals. They are well-arranged without any defect inside the enclosure.
   6. Mechanical check for all components inside the control panel.
   7. All required power supplies are provided according to the approved submittal and shop drawings.

E. Function tests should be carried out for the system to ensure the system is working according to project specifications.

   1. Check when the system is in operation that all channels' signals are clear at each outlet without any distortion.
   2. The signal level for the system should be checked and programmed to the approval level without any distortion.
   3. The trial pictures of various channels at the amplifier output showing correct operation. Amplifier gain to be adjusted for clarity.

4. Check that all channels can be seen in each outlet.

F. Final approval for the system is subject to the followings:

1. Successfully testing and commissioning of the system.
2. The performance of the equipment/system should achieve a minimum signal level of 65 dB for each outlet.

Operation and maintenance (O&M) manual for MATV System

The operation and maintenance manual should contain the following.

1. Description of the project system.
2. How to use the manual.
3. Troubleshooting guide.
4. Safety instruction at work.
5. Start-up and shutdown procedure.
6. An original catalogue highlighting the material used with part number.
7. As built drawing including schematic and riser diagram.
8. List of spare parts.
9. Procedure for routine and preventive maintenance for monthly/1 year/5 years.
10. Warranty certificate for 2 years (as per project requirement).
11. Testing and commissioning certificates/reports.
12. Emergency information.
13. Manufacturer/supplier information and contact details.
14. Overview and method of operation (sequence of operation).
15. Training for the operation and maintenance team. (Training can be done by submitting a training manual for the consultant and client approval).
16. System configuration and programming.
17. All guidelines for using all equipment and components.

Parking Management System

Approval for Material Submittal for Parking Management System

The consultant should follow the contractor in arranging periodical meetings between the electrical engineer and the supplier of the parking management system. The meeting should discuss and clarify the mode of operation, material submittal and work progress. The contractor should be instructed by the consultant to provide the followings to get the consultant's approval for the material submittal:

1. The selected brand or manufacturer is listed in the consultant's approved manufacturer list.
2. A brief description of parking management and sequence of operation.
3. Compliance statement to be submitted. The supplier should clearly show his compliance with the specifications, and non-compliant sections should be highlighted with justifications.
4. The submittal should include the original catalogue highlighting the selected components and the technical points complied to the project specifications. Also, to include the sequence and philosophy of operation.
5. Valid supplier trade license with expiry date and Ministry of Trade certification as a legal supplier of the material in the submittal.
6. All indicator lamps must be low voltage with a built-in transformer, or lamps must be L.E.D. type.
7. Details of control panel and all other accessories.
8. The degree of protection (I.P. rating) should be according to the specifications and the location of installation.
9. All accessories, wiring, etc., should be part of the submittal.
10. Loading calculations and wire selection should be submitted.
11. Riser diagram for a parking management system with full details, including all related components.
12. Each circuit (zone) should have 10% spare for future additional sensors or as per project specifications.
13. Wiring diagram for control panel including terminal block showing details for all ingoing and outgoing wires to the control panel.
14. The control panel should be capable of interfacing with B.M.S. and other systems according to the project specifications.
15. The control panel should have an Indication lamp on the panel indicating the availability of the power.
16. Identification using engraved labels and ferrules to all cables, control wiring, and accessories to be included within the submittal.

17. System earthing should be submitted based on the manufacturer's requirements.
18. Communication compatibility with B.M.S. and all interfaced systems.

Workshop drawings for Parking Management System.

The consultant should instruct the contractor to prepare the workshop and coordination drawings for the Parking Management System to be submitted for the consultant's approval. The followings should be considered in the drawings:

1. Workshop and coordination drawings for the parking management system should be part of lighting drawings clarifying the indication lamp locations, Number of loops/zones, main control panel, and reference of power supply in coordination with all adjacent services around.
2. Coordination drawings and details for the control room/ Security room that contain the parking management system panel showing the Main Control Panel location, monitoring panel, dimensions, and connection to B.M.S. (if applicable).
3. Workshop drawings to be signed by the suppliers of the material. The signature gives indication that the supplier specialist is accepting the installation and the installation details are matching the supplied system.
4. Flow chart for all activities related to the system showing all interlocks and interfaces with other systems and scope of all involved contractors.
5. Block diagram/ Schematic diagram for all activities of the system and other interlocks.
6. The power supply for the control panel and main control equipment should be fed from separate sections with a breaker and ELCB (the circuit reference for the power supply should be marked on the drawing). The power supply is preferred from emergency D.B. if available.
7. Power supply outlet for parking management system should be with fused switch. Power supply should be fed from separate Section, ELCB, M.C.B. through emergency D.B. if available.
8. Drawing of the Network between all components and control panels.
9. Ensure that the number of points in each Zone/ circuit is studied in terms of circuit capacity, loading, and voltage drop.
10. Ensure that each grouping (zone/circuit) has been studied and coordinated with (Architectural work and parking layout).
11. Details of earthing for the system should be done as per supplier recommendation.
12. Location of all points should be coordinated with all services inside parking like lighting fixture, A/C duct cable tray, parking layout. Etc.

13. Mounting height of all devices should be shown in approved workshop drawing or approved mounting height schedule. The height should allow the highest car parking beneath the sensor. (Supplier specialist should be consulted and he should sign on the shop drawings confirming it is correct).

Installations of Parking Management System

The consultant should follow the contractor to ensure the installation of the following activities and devices for the parking management system:

1. The contractor should submit an Installation method statement to the consultant for approval.
2. Before installation, ensure all materials are as approved in the submittal and without any damages.
3. Activity chart, work progress and look ahead program of work for Parking Management System clarifying the steps for executing the work such as (Defining the scope of work for involved contractors, Civil work, Electrical Work, Testing and Commissioning) with monthly updates.
4. The contractor should use appropriate back boxes for all components in the installed system.
5. Ensure installing all conduits and wires between all points, control panel, in the control room ...etc.
6. Proper protection for all devices shall be provided during installation work.
7. The final location of all points should be according to approved workshop and coordination drawings.
8. Components Identification for the system (devices, wiring, junction box ...etc.) should be made using engraved labels and ferrules.
9. Earthing for the system should be done according to supplier/manufacturer recommendations.
10. A wiring diagram for the control panel shall be provided inside the control panel.
11. All wiring for control and power to be as per standards, project specifications and authority regulations. Voltage drop calculation shall be considered when wire sizing.
12. A backup power supply (battery) shall be provided inside the control panel.
13. The mounting height of devices should be according to the workshop/ coordination drawings or approved mounting height schedule.
14. Terminations are robustly done on the correct terminals for all devices, control panel, ... etc.
15. Supplier, Contractor's Site Engineer should daily and continuously monitor the installation activity till completion. Periodical progress report showing site status, requirements ...etc. Shall be submitted by the contractor.
16. Break down for all item related to the system shall be proposed by contractor for billing purpose; these broken-down percentages should be agreed by all parties.

17. Segregation between control data cabling and power wiring should be maintained for safety and noise immunity.
18. All control wires and power cables shall be with extra length for easy connection to the equipment terminals.

Testing and commissioning of Parking Management System.

A. The consultant should instruct the Contractor to complete all pending or delayed work related to the (sensors with built-in electronic pilot light, power cable, communication cable ...etc.). Testing and commissioning method statements should be submitted by the contractor for the consultant's approval, including standard readings and successful test criteria.

B. Complete the wiring megger test before starting any termination for the system components.

C. Confirmation letter from the supplier stating that the parking management installation is completed and is ready for operation and testing.

D. Visual inspection and check for the system confirming that proper installation and termination are completed and confirm the followings:

   1. Installation of the main control panel, all accessories and sensors...etc. are according to the approved workshop and coordination drawings.
   2. The power supply is connected as per system requirements.
   3. The main control panel, sensor ...etc. are identified and labelled using engraved labels and ferrules.
   4. The wiring diagram is provided inside the control panel.
   5. All components are procured according to the approved material submittal and installed according to the approved workshop and coordination drawings.
   6. Mechanical checking for all components inside the control panel.

E. Function tests should be carried out for the system to ensure the system is working as per project specifications. Make sure when power supply is off the system still in operation by using the battery power supply.

F. Final approval for the system is subject to successful testing and commissioning (could be by a third party; refer to the project specifications).

Operation and Maintenance Manual for Parking Management System

The consultant should follow the contractor in submitting the O&M manuals for the Parking Management system, including the following requirements.

1. Description of the project system.
2. How to use the manual.
3. Troubleshooting guide.
4. Safety instruction at work.
5. Start-up and shutdown procedure.
6. An original catalogue highlighting the used material with part number.
7. As built drawing including schematic riser diagram/block diagram.
8. List of spare parts.
9. Procedure for routine and preventive maintenance for monthly/1 year/5 years.
10. Guidelines for using all devices/controllers.
11. Warranty certificate for 2 years (as per project specifications).
12. Testing and commissioning certificates/reports.
13. Emergency information.
14. Manufacturer/supplier information and contact details.
15. Overview and method of operation (sequence of operation).
16. Training for the security, operation, and maintenance team. (Training can be done by submitting a training manual for the consultant and client approval).
17. System programming and configurations.

Uninterruptable Power Supply (U.P.S.) System

Approval for Material Submittal for U.P.S. System

The consultant should follow the contractor in arranging periodical meetings between the electrical engineer and the suppliers of the U.P.S. system and other related or interfaced systems like B.M.S., switch gear, hydrogen sensing system, etc. The meeting should discuss and clarify the modes of operation during fire or power supply interruption, material submittal, and progress of work. The contractor should be instructed by the consultant to provide the followings to get the consultant's approval for the material submittal:

1. The selected brand or manufacturer is listed in the consultant's approved manufacturer list.
2. An original catalogue highlighting all technical points in the specifications.
3. Description for U.P.S., sequence, and mode of operation.
4. Flow chart for all D.B.s connected to U.P.S. system clarifying D.B.'s location and details.
5. Compliance statement to be submitted. The supplier should clearly show his compliance to the project's specifications, non-complied sections should be highlighted with justifications.
6. Valid supplier trade license with expiry date, and ministry of trade certification as a legal supplier of the material in the submittal.
7. All indicator lamps must be low voltage and have a built-in transformer or L.E.D.
8. Batteries submittal showing Autonomy and sizing calculations based on project specifications requirements to be submitted attached with U.P.S. submittal for consultant's approval.
9. The supplier should clarify whether the battery room requires a hydrogen sensing system that controls exhaust fans (based on battery type).
10. U.P.S. entry and outgoing breakers should be coordinated with supply and load breakers.
11. Block diagram for U.P.S. system clarifying in detail all components of the cabinet like a filter, battery charger, static bypass switch, input stage, output stage, inverter, display board, control board, interphase with other systems, main power supply, etc.
12. Consultant and Authority approval for the following before ordering the U.P.S. for correct sizing:
    a. Schematic and single-line diagram for U.P.S. load.
    b. Workshop drawing for the load fed by the U.P.S.
    c. Approval for equipment load fed by the U.P.S. (actual load).
    d. Approval for operation and the required time delay after power failure.
    e. The approved list of all SMDBs and D.B.s will be fed by U.P.S.

13. Details of the earthing system for UPS to be included in the submittal.
14. Communication protocol should be compatible with other interfaced systems and B.M.S.
15. Software for Setup, programming and configurations.

Workshop drawings for U.P.S.

The consultant should instruct the contractor to prepare the workshop and coordination drawings for the Uninterruptable Power Supply (U.P.S.) The system should be submitted for the consultant's approval. The followings should be considered in the drawings:

1. Workshop and coordination drawings for the U.P.S. system should show the location of the cabinet, D.B.s, and all circuits or the distribution system connected to U.P.S. outputs using correct wire and cable sizing.
2. Coordination drawing and details for U.P.S. and batteries' room showing the followings:
    a. Location of U.P.S.
    b. distribution cabinet,
    c. monitoring panel,
    d. battery room arrangement,
    e. Hydrogen sensing system and exhaust fans in battery room
    f. connection to B.M.S.
    g. Lighting fixture and switch
    h. All power outlets.
    i. Emergency light
    j. A/C System
    k. Fire Alarm system (detectors)
    l. Firefighting system.
3. Riser diagram for U.P.S. system clarifying in detail all components, circuits, power supply, size of wires, rectifier, charger, inverter, static switch, manual bypass switch, battery bank ...etc.
4. Shop drawings to be signed by the suppliers. The signature gives indication that the supplier specialist is accepting the installation and coordination drawings, and the installation details are matching the supplied system.
5. U.P.S. supply should be from a dedicated MCB+ ELCB or MCCB, depending on U.P.S. size and installation location in the project distribution system.

Installation of U.P.S.

The consultant should follow and instruct the contractor in the installation stage of the U.P.S. system to check the followings:

1. The contractor should submit an installation method statement for the system for the consultant's approval.
2. Contractor should request the consultant to inspect the received material to the site and ensuring passed M.I.R. (Material Inspection Request) and approved by consultant.
3. Input, output, load and battery cables and wires are connected properly with correct colour coding. The connection of the wires and cables is done according to the project design and supplier specialist guidance.
4. Proper protection for all wiring, U.P.S. and distribution systems should be provided during storage and installation.
5. The final location of cabinets, distribution panels and other accessories should be as per approved workshop and coordination drawings, taking into consideration the followings:
   a. Proper supporting system for the equipment on the foundation.
   b. Enough space and clearance around the panel for maintenance, ventilation, and component replacement.
6. Identification for the system (control panel, cabinet, battery banks ...etc.) shall be provided using engraved labels and ferrules.
7. Ensure that there is proper ventilation/air conditioning in the U.P.S. and Battery rooms based on manufacturer recommendations.
8. Ensure the power supply for the U.P.S. is fed from a separate breaker/section / ELCB.
9. U.P.S. earthing should be done according to the supplier's recommendation.
10. Connection of the monitoring panel and connection to B.M.S.
11. Break down for all items related to the system shall be proposed by contractor for billing purpose; these broken-down percentages should be agreed by all parties.
12. Power and control cables and wires are selected according to the standard, regulations and specifications taking into consideration the voltage drop calculation.
13. Termination is robust and correct for all devices, control panels, batteries banks, etc.
14. Supplier specialist confirmation regarding the U.P.S. and battery room ventilation confirming that it is provided according to the manufacturer's recommendations.
15. Based on battery type, battery room may require hydrogen sensing system that control exhaust fans (coordination with mechanical engineers).

16. Supplier/ Contractor's Site Engineer should monitor daily and continuously the installation activity till completion. Periodical progress report showing site status, requirements ...etc. Shall be submitted by the contractor.
17. Segregation between control data cabling and power wiring should be maintained for safety and noise immunity.
18. All control wires and power cables shall be with extra length for connection to the equipment terminals.

Battery room Safety System

For hydrogen-generating batteries, battery room safety system should be designed and procured complying with the followings:

1. Batteries should be installed in separate rooms isolated from other systems or equipment and should have sufficient space according to the standards. The floor should be anti-acid protected.
2. An air conditioning system is should provided to maintain room temperature at 25C and to be above the due point in order to prevent condensation of moisture.
3. Ventilation system of capacity of 5 times room air to be changed per hour. Extract fan to extract air from top of the room, the extract fan should be hazardous proof. The ventilation should be sufficient to keep the generated hydrogen gas below 1% limit, then the space can be considered non-hazardous are according to NFPA 70.
4. The air extractor should be controlled by an airflow switch connected to a rate charging switch in the system control panel to stop the battery charging when it is not working for any reason.
5. Hydrogen gas sensor system with alarm horn to be fixed in the room connected to the fire alarm control panel, and to the high-rate charge switch in the system controller to give alarm and indication in the FACP, and to stop charging the batteries when hydrogen exceeds the accepted limit.
6. Lighting fixtures and gas detector should be hazardous proof type; rest of electrical installations at the lower part of the area could be of non-hazardous proof type, as per note 3.
7. NO SMOKING sign should be fixed inside the room and on the access door.
8. Firefighting extinguishers of powder and sand should be provided inside the room.
9. The contractor must ensure that all safety requirements are in place and accessible in suitably clean and dry conditions in accordance with the project safety plan.

Testing and commissioning of U.P.S.

A. Before starting the testing of the U.P.S. system, the consultant should instruct the contractor to complete all pending and delayed work related to the U.P.S. system. Testing and commissioning method Statement to be submitted by the contractor for the consultant approval including standard readings and successful test criteria.

B. A megger test for incoming and outgoing cables and wires shall be done before starting any termination of the system.

C. A confirmation letter from the supplier is needed that the U.P.S. system installation is completed and ready for operation and testing.

D. Visual inspection and check for the system to ensure that proper fixing, installation, and termination are done for the followings:

1) The main equipment inside the U.P.S. room and control room.
2) Location of all cabinets and other accessories according to the approved workshop and coordination drawings.
3) The power supply has been connected according to system requirements.
4) The main control panel, monitors, cabinet, etc... are identified and labelled using engraved labels and ferrules.
5) Testing and checking the input and output cable insulation resistance by Megger.
6) The correct value for input voltage, frequency, and input phase sequence.
7) The proper polarity of the battery banks, input, output and control cables and rating of all fuses.
8) Ensure that all components are installed and the battery connection is correct according to the approved submittal and workshop drawings.
9) Mechanical checking for all components inside the U.P.S.

E. Function Test by Energizing the input supply and verifying the following:

1) Switch on the U.P.S. and verify that:
   a) No fault or alarm.
   b) All indicators and fuses are in operation.
2) Switch on the battery breaker and verify that:
   a) Battery voltage is ok.
   b) The battery is being charged.
3) Connect the minimum load to the bypass supply and verify for synchronization. If ok, then transfer the load to U.P.S. output and check the followings:

a) U.P.S. output voltage, phase, and frequency.

b) No alarm was present.

4) Connect the full load to the U.P.S. and verify the followings:

1. Simulate fault condition, ensuring proper load transfer to bypass mode.
2. All load breakers are switched on.

F. Final approval for the system is subject to successful testing and commissioning.

Operation and Maintenance Manual for U.P.S.

The operation and maintenance manuals should include the followings:

1. Description of the project system.
2. How to use the manual.
3. All guidelines for using devices and controllers.
4. Troubleshooting guide.
5. Safety instruction at work.
6. Start-up and shutdown procedure.
7. An original catalogue highlighting the material used with part numbers.
8. As-built drawings, including a schematic riser diagram for U.P.S. distribution power, clarify the interfacing with other systems.
9. List of spare parts and special tools.
10. Procedure for routine maintenance for monthly/1 year/5 years (as per project specification).
11. Warranty certificate for 2 years (as per project specification).
12. Testing and commissioning certificates/reports.
13. Emergency information.
14. Manufacturer/supplier information and contact details.
15. Configurations, modes and sequence of operation, settings, software, and programming.
16. Training for security, operation, and maintenance teams. ((This can be done by submitting a separate training manual including training details and session content).

Nurse Calling System

Approval of the material submittal for the Nurse Calling System.

The consultant should follow the contractor to arrange periodical meetings between the Electrical engineer, the supplier of the Nurse Calling System and all other suppliers of other systems or components connected or interfaced to the N.C.S. system like the nursing station monitor unit, over-door indicator unit … etc., to discuss and clarify the mode of operation, material submittal work progress etc. The contractor should provide the followings for the consultant's approval for the material submittal:

1. The selected brand or manufacturer is listed in the consultant's approved manufacturer list.
2. A brief description of the N.C.S. system and sequence of operation.
3. Compliance statement to be submitted. The supplier should clearly show his compliance to the specifications, non-complied sections should also be highlighted with justifications.
4. Flow chart for all components and devices, including the location, type, power supply … etc. and any other interfaced system with N.C.S. according to the project specification and authority requirements.
5. The submittal should include the original catalogue highlighting the selected components and the technical points complied to the project specifications. Also, to include the sequence and philosophy of operation.
6. Valid supplier trade license with expiry date, and ministry of trade certification as a legal supplier of the material in the submittal.
7. Block diagram for N.C.S. system showing all details including all activities connected to the system like (power supply reference, over door indicator, patient call unit, in room Nurse response/call rest unit etc…) and all other interfaced systems.
8. Control panel should be capable of interfacing with all related systems based on project specifications, authority and client requirements.
9. All indicator lamps must be either low voltage with built-in transformers or L.E.D. lamps.
10. Identification and numbering for all devices wiring, junction box and other devices shall be done by using engraved labels and ferrules. The identification details should be part of the material submittal.
11. Degree of protection of the system should be according to the project specifications and the location the system will be used.
12. All accessories and cables related to the system should be included in the submittal with details and catalogues.
13. Earthing system requirements should be provided within the submittal.
14. Communication compatibility with B.M.S. and all interfaced systems (if applicable by project specifications).

Workshop Drawings for Nurse Calling System

The consultant should instruct the contractor to prepare the workshop and coordination drawings for the Nurse Calling System to be submitted for the consultant's approval. The followings should be considered in the drawings:

1. Wiring diagram for the Control Panel, nursing station monitor unit... etc., including terminal block showing details for all ingoing and outgoing wires to the control panel or nursing station monitor unit.
2. Workshop and coordination drawings for the N.C.S. system should be submitted with power drawings, including all components, devices, mounting height and installation details coordinated with other M.E.P. services, etc... The drawings should show the interfaced systems, services and components which are connected or interfaced to N.C.S.
3. Coordination drawing and details for the control, monitoring, and patient rooms showing location of main nursing station, monitoring unit of the system and system components in patient's rooms coordinated with furniture layout and all other services.
4. Shop drawings to be signed by the suppliers of the material. The signature gives indication that the supplier specialist is accepting the installation and the installation details are matching the supplied system.
5. Power supply for the N.C.S. Control Panel should be according to supplier specialist requirement, fed from separate sections (ELCB and M.C.B.) using fuse connected outlet with indication lamp, preferably from emergency D.B. if available.

Installation of Nurse Calling System

The consultant should follow and instruct the contractor in the installation stage of N.C.S. and its interfaced systems. Contractors to check and ensure the followings:

1. The contractor should submit a method statement for installation activities for the whole system with all related interfaces and components.
2. Activity chart, work progress and look ahead program of work for N.C.S. system clarifying steps for executing the work (civil, electrical, finishing work, furniture layout testing and commissioning) defining the scope of work for each involved contractor with monthly updated.
3. Ensure all used materials match the approved material submittal.
4. Ensure that the contractor is using the correct-sized back boxes for all components.
5. Ensure installing all conduits and wires between the nursing station monitoring unit and other systems' equipment.

6. Proper protection for all devices, wires, junction boxes...etc. should be provided during installation work.
7. The mounting height of all devices should be according to the approved workshop/ coordination drawings or approved mounting height schedule.
8. Ensure approval for the philosophy of operation of the system by the consultant before installation.
9. All control wires and power cables should be installed as per specifications, standards and tested for continuity and insulation tests; voltage drop calculation shall be considered.
10. Identification and numbering of all devices, wiring, junction box and other devices should be done by using engraved label and ferrules.
11. N.C.S. earthing should be installed according to supplier specialist and manufacturer recommendations.
12. Ensure that all termination is robustly done on the correct terminals for all devices and control panel, etc....
13. Supplier, Contractor's Site Engineer should daily and continuously monitor the installation activity till completion. Periodical progress report showing site status, requirements ...etc. Shall be submitted by the contractor.
14. Segregation between control data cabling and power wiring should be maintained for safety and noise immunity.
15. All control wires and power cables shall be with extra length for connection to the equipment.
16. Break down for all items related to the system shall be proposed by contractor for billing purpose; these broken-down percentages should be agreed by all parties.

Testing and commissioning of Nurse Calling System.

1. The consultant should instruct the Contractor to complete all pending or delayed work related to the N.C.S. and interfaced systems. Testing and commissioning method statements should be submitted by the contractor for the consultant's approval, including standard readings and successful test criteria.
2. Complete the wiring megger test before starting any termination for the system components.
3. A confirmation letter from the supplier is needed confirming that the N.C.S. System installation is completed and ready for operation and testing.
4. Visual inspection and check for the system to ensure that proper installation and termination are done for the followings:
    a) Main control panel, nursing station monitor unit, power supply, over door indicator unit, inter-phase connection etc...

b) System earthing and power supply connection are completed according to the standard and supplier specialist/ manufacturer requirements.

c) Main control panel is dressed neatly, wiring diagram is included in the panel. Incoming and outgoing wires and cables to terminal block are terminated with engraved identification labels and ferrules.

d) Ensure that all components are selected and installed according to the approved material submittal and workshop/ coordination drawings.

5. Function tests should be carried out on the system after installation and connection are completed. The test should include testing the operation in case of power failure.

6. Final approval for the system subject to Successful testing and commissioning of the system by third party (based on contract specifications).

Operation and Maintenance Manual for Nurse Calling System.

The consultant should follow the contractor to submit the O&M manuals for the N.C.S. system, including the following requirements:

1. How to use the manual.
2. Description of the project system.
3. Safety instruction at work.
4. Start-up and shutdown procedure.
5. Original catalogues for all materials highlighting the used material.
6. All guidelines for using system devices and controllers.
7. Overview and method of operation (sequence of operation).
8. Emergency information.
9. List of spare parts.
10. Troubleshooting guide (for emergency breakdown).
11. Procedure for routine and preventive maintenance for monthly/1 year/5 years.
12. Warranty certificate for 2 years (as per project specifications).
13. Testing and commissioning certificates/reports.
14. Manufacturer/supplier information and contact details.
15. Training for the security, operation, and maintenance team. (Based on the project, it may be required to submit a training manual for approval of training content).

16. System configurations, programming, settings, and the used software, if applicable.
17. Original catalogue with highlights on the material used.
18. As built drawing including schematic riser diagram/block diagram.

Central Clock and Bell System

Approval of material submittal for Central Clock and Bell System

The consultant should follow the contractor to arrange periodical meetings between the Electrical engineer and the suppliers of the Central Clock and Bell System and all other interfaced and interlocked systems' suppliers like B.M.S. and all other systems that need connection to the common clock system to discuss and clarify the mode of operation, material submittal requirements, work progress ..etc. The contractor should provide the followings for the consultant's approval for the material submittal:

1. The selected brand or manufacturer is listed in the consultant's approved manufacturer list.
2. A brief description of the central clock and bell system sequence of operation should be submitted for approval.
3. Compliance statement to be submitted. The supplier should clearly show his compliance to the specifications, non-complied sections should be highlighted with justifications.
4. The submittal should include the original catalogue highlighting the selected components and the technical points complied to the project specifications. Also, to include the philosophy of operation.
5. Valid supplier trade license with expiry date, and ministry of trade certification as a legal supplier of the material in the submittal.
6. The control panel should be able to interface with all related systems according to the project specifications.
7. All indicator lamps must be low voltage with a built-in transformer, or lamps must be L.E.D. type.
8. All accessories and required components for installation should be included in the submittal.
9. Block diagram for central clock and bell system including details for all activities connected/ related to the system like power supply reference, clock, bell connection to B.M.S., ..etc.
10. System earthing should be clarified in the submittal based on manufacturer recommendations.
11. Wiring diagram for control panel including terminal block showing details for all ingoing and outgoing wires to the control panel.
12. Spares should be considered in the installed system for future expansion. (Refer to the project's specifications).
13. Communication compatibility with B.M.S. and all interfaced systems.

Workshop drawings for Central Clock and Bell System

The consultant should instruct the contractor to prepare the workshop and coordination drawings for the

Central Clock and Bell System to be submitted for the consultant's approval, the followings should be considered in the drawings:

1. Workshop drawing for the central clock and bell system should be submitted with the power drawings clearly indicating location of the main control panel, clock, bell, G.P.S. satellite time signal, junction box, mounting height etc... showing the routing of all conduits to power, control, and interfaces with other systems (B.M.S., etc.) and coordinated with all other systems and services.
2. Control room coordination drawings with coordination details showing location of all panels related to the system and connection between panels and network or connection to other interfaced systems like B.M.S. (if any).
3. Shop drawings to be signed by the suppliers of the material. The signature gives indication that the supplier specialist is accepting the installation and installation details are matching the supplied system.
4. Flow chart for all activities related to the system showing all interlocks and interfaces with other systems and scope of all involved contractors.
5. Block diagram/ Schematic diagram for all activities of the system and other interlocks.
6. Power supply for the system should be from separate Section/ELCB/MCB and from emergency D.B. if available.
7. System earthing should be shown based on manufacturer requirements.
8. Location of system components and clocks to be coordinated with furniture layout and other services.
9. The mounting height of all devices should be shown in the approved workshop drawing or approved mounting height schedule.

Installation of Central Clock and Bell System.

The consultant should follow and instruct the contractor in the installation stage of the central clock and bell systems with all related and interfaced systems to check and ensure the followings:

1. Supplier should submit installation method statement for consultant's approval for the whole system with all related interfaces and components.
2. The contractor should ensure the consultant's approval for the philosophy of operation of the system before installation on the site.
3. Before installation, ensure all materials are as approved in the submittal and without any damages.
4. The contractor should use appropriate back boxes for all components in the installed system.
5. Ensure installing all conduits and wires between B.M.S. and interfaced systems, power supply (isolator and D.B.) control room etc...

6. Proper protection for all wiring and devices shall be provided during installation.
7. Mounting height and location of all clocks, bells and other accessories shall be as per approved workshop and coordination drawings or approved mounting height schedule.
8. Identification and numbering for system components (isolator, wiring, junction box …etc.) shall be provided using engraved labels and ferrules.
9. Earthing of the system shall be according to the supplier's or manufacturer specialists' recommendation.
10. All control wires and power cables shall be installed according to the project's specifications and standards, voltage drop calculation shall be considered.
11. Ensure that all termination is robustly done on the correct terminal for all devices, control panels, etc…
12. Wiring diagram shall be provided inside the control panel.
13. Supplier, Contractor's Site Engineer should daily and continuously monitor the installation activity till completion. Periodical progress report showing site status, requirements ...etc. Shall be submitted by the contractor.
14. Segregation between control data cabling and power wiring should be maintained for safety and noise immunity.
15. All control wires and power cables shall be with extra length for easy connection to the terminal equipment.
16. Activity chart, work progress and look ahead program of work for central clock and bell system clarifying all steps for executing the work such as civil work, electrical work, testing and commissioning, finishing work and defining the scope of work for the involved contractors with monthly update.
17. Breakdown for all items related to the system shall be proposed by the contractor for billing purposes; these broken-down percentages should be agreed upon by all parties.

Operation and Maintenance (O&M) Manual for Central Clock and Bell System.

The consultant should follow the contractor to submit the O&M manuals for the Intercom system, including the following requirements:

1. Description of the project system.
2. How to use the manual.
3. Safety instruction at work.
4. Start-up and shutdown procedure.
5. Overview and method of operation (sequence of operation).
6. Procedure for routine and preventive maintenance for monthly/1 year/5 years.

7. All guidelines for using devices and controllers for the control panel.
8. System settings, programming, configurations, and the used software, if applicable.
9. Troubleshooting guide (for emergency breakdown).
10. Original catalogue with highlights on the material used.
11. As built drawing including wiring diagram for the control panel and block diagram.
12. List of spare parts.
13. Warranty certificate for 2 years (as per project requirement).
14. Testing and commissioning certificates/reports.
15. Emergency information.
16. Manufacturer/supplier information and contact details.
17. Training for security, operation, and maintenance teams. (Training can be done by submitting a training manual for the consultant and client approval).

Testing and commissioning for the Central Clock and Bell system

1. Testing and commissioning are to be done by a third party (refer to the project specifications).
2. Final approval for the system is subject to successful testing and commissioning.

Privat Branch Exchange (PBX)

Approval of material submittal for Private Branch Exchange System

The consultant should follow the contractor in arranging periodical meetings between the electrical engineer and the supplier of the PBX system. The meeting should discuss and clarify the mode of operation and any requirement of the authority or service provider. The consultant should instruct the contractor to provide the followings to get the consultant's and then the authority's approval for the material submittal:

1. The selected brand or manufacturer is listed in the consultant's approved manufacturer list (the brand may also require authority approval).
2. A brief description of the system and sequence of operation.
3. Compliance statement to be submitted. The supplier should clearly show his compliance to the specifications, non-complied sections to be highlighted with justifications.
4. The submittal should include the original catalogue highlighting the selected components and the technical points complied to the project specifications. Also, to include the sequence and philosophy of operation.
5. Valid supplier trade license with expiry date, and ministry of trade certification as a legal supplier of the material in the submittal.
6. Flow chart for all telephone point connected to PBX system and related or interfaced systems according to the project specifications clarifying location and details of PBX.
7. Block diagram for PBX system showing all systems details, including all activities connected/related to the system like power supply reference, connection to B.M.S. in control room …etc.
8. All indicator lamps must be low voltage with a built-in transformer, or lamps must be L.E.D. type.
9. System earthing should be done based on manufacturer and authority requirements.
10. All accessories, back boxes, wiring, etc., should be part of the submittal.
11. The degree of protection (I.P. rating) should be according to the specifications and the location of installation.
12. Spares should be considered in the system for future expansion (Refer to the project's specifications).
13. Check with the manufacturer if the PBX requires cooling or ventilation to avoid overheating the equipment.
14. Communication compatibility with B.M.S. and all interfaced systems.

Workshop Drawings for PBX

The consultant should instruct the contractor to prepare the workshop and coordination drawings for the Private Branch Exchange System to be submitted for the consultant's and the authority's approvals. The followings should be considered in the drawings:

1. Workshop and coordination drawings for telephone system (PBX) clarifying locations and mounting height of main control panel, M.D.F., I.D.F., telephone point, power supply reference …etc. Drawings to be approved by Authority.
2. Telephone rooms shop and coordination drawings with full details showing location of Main Panels, intermediate panels, PBX, and connections to other interfaced systems.
3. Based on the project specifications, contractor should check if the PBX requires U.P.S. for redundant supply, if U.P.S. Is required, then to show the U.P.S. installation details and dimensions in the coordination drawings.
4. A riser diagram for telephone system details, including all activities connected to the system (circuit/ power supply details, size of wiring, outlet locations, etc.)
5. Shop drawings should be signed by the supplier of the material. The signature give indication that the supplier specialist is accepting the installation and the installation details are matching the supplied system.
6. Flow chart for all activities related to the system showing all interlocks and interfaces with other systems and scope of all involved contractors.
7. Power supply for PBX system should be from separate section/ELCB/MCB. The supply should be from emergency D.B. if available by using a fused switch or cable outlet with an indicator lamp.
8. Details of PABX earthing are based on the supplier specialist's recommendations.

Installation of PBX

The consultant should follow the contractor to ensure the installation of the following activities and devices for the PBX system with all related and interfaced systems and components, the contractor should check and ensure the following:

1. The contractor should submit an installation method statement for the whole system with all related interfaces and components.
2. Ensure that all conduits and wires are installed between all telephone points, main and intermediate panels.
3. Before fixing components, ensure all received materials are as per approved material submittal and authority requirements.

4. Ensure that the contractor uses proper back boxes for the installed components.
5. Ensure that all required conduits between back boxes and the required containment for all systems are installed.
6. IF UPS is required for the PBX, the contractor should ensure installing the U.P.S. with the PBX to certify installation completion.
7. Proper protection for all devices shall be provided during the system's installation stage.
8. Identification and numbering for telephone outlet, wiring, junction boxes and other devices shall be done using engraved labels and ferrules. Telephone numbers table to be provided and to be handed over to the client in project handing over.
9. Ensure that all terminations are done correctly and robustly on the correct terminals for all devices, control panels, etc...
10. Final location of telephone point should be according to approved workshop and coordination drawings. It is advisable to be near power supply outlet.
11. All wiring for control and power should be according to the standards, project specifications, and authority regulations.
12. Mounting height and location of all telephone points and control panels shall be as per approved coordination drawings or approved mounting height schedule.
13. Activity chart, work progress and look ahead program of work for PBX system clarifying steps for executing the work (civil work, electrical work, finishing work, testing, and commissioning). Defining the scope of work for all involved contractors with monthly updates).
14. System earthing shall be according to supplier/manufacturer recommendation.
15. Segregation between Telephone/ data cabling and power wiring should be maintained for safety and noise immunity.
16. All telephone/ data wires and power cables shall be of extra length for connection to the terminal equipment.
17. Supplier, Contractor's Site Engineer should daily and continuously monitor the installation activity till completion. Periodical progress report showing site status, requirements ...etc. Shall be submitted by the contractor.
18. Break down for all item related to the system shall be proposed by contractor for billing purpose; these broken-down percentages should be agreed by all parties.

Testing and Commissioning of PBX.

A. The consultant should instruct the Contractor to complete all pending or delayed work related to the system (telephone point/control panel ...etc.). T&C Method Statements to be submitted by the contractor for the consultant's approval including standard readings and the successful test criteria.

B. Complete the wiring test before starting any termination for the system components.

C. A confirmation letter from the supplier is needed confirming that the PBX system installation is completed and ready for operation and testing.

D. Visual inspection and check for the system to ensure that proper installation and termination are done for the followings:

   1. The power supply has been connected according to system requirements.
   2. The Main Control Panel, telephone, junction box ...etc. are identified and labelled by using engraved labels and ferrules.

E. Function test to ensure the system is working according to project's specification. The system to be tested and telephone numbering of the telephone outlets to be checked in addition to check the sound and operation of switch board in the reception/control room.

F. Final approval for the system subject to successfully testing and commissioning of the system. Testing and commissioning should be done by third party (Refer to the project's specifications).

Operation and Maintenance Manuals (O&M) for PBX

The consultant should follow the contractor in submitting the O&M manuals for the PBX system, including the following requirements.

1. Description of the project system.
2. How to use the manual.
3. Troubleshooting guide for the system.
4. Safety instruction at work.
5. Start-up and shutdown procedure.
6. An original catalogue highlighting the used material with part number.
7. As-built drawings including wiring and block diagrams.
8. List of spare parts.
9. Procedure for routine and preventive maintenance for monthly/1 year/5 years.

10. System programming, settings, configurations, and the used software, if applicable.
11. Warranty certificate of the project (as per project requirement).
12. Testing and commissioning certificates/reports.
13. Emergency information.
14. Manufacturer/supplier information and contact details.
15. Overview and way of operation (sequence of operation).
16. Training for the security, operation, and maintenance team. ((This can be done by submitting a separate training manual including training details and session content).
17. All guidelines for the operation of devices and controllers inside the control room.

Smart Lighting and Dimming Control System

Approval of material submittal for Smart Lighting and Dimming Control System

The consultant should follow the contractor to arrange periodical meetings between the Electrical engineer and the Supplier of S.L. and D.C. systems and related interfaced systems suppliers like B.M.S… etc. The meeting should discuss and clarify the mode of operation, interlocks, interfaces with other systems, material submittal and work progress. The contractor should be instructed by the consultant to provide the followings to get the consultant's approval for the material submittal:

1. The selected brand or manufacturer is listed in the consultant's approved manufacturer list.
2. A brief description of S.L. and D.C. and the sequence of operation.
3. Compliance statement to be submitted. The supplier should clearly show his compliance to the specifications, non-complied sections should be highlighted with justifications.
4. Block diagram for S.L. and D.C. system showing all system's details, including all activities connected to the system like connection with F.A. system, emergency lighting, B.M.S., access control, control panel, … etc.
5. The submittal should include the original catalogue highlighting the selected components and the technical points complied to the project specifications. Also, to include the sequence and philosophy of operation.
6. Valid supplier trade license with expiry date, and ministry of trade certification as a legal supplier of the material, should be included in the submittal.
7. Wiring and cables should be part of the submittal.
8. Engraved labelling should be part of the submittal.
9. Motion detector data sheet and specifications (Type and timing) shall be submitted for approval.
10. Override switches' technical details to be submitted. location to be finalized coordinated with all other services, architectural work and as recommended by supplier specialist.
11. List all equipment related to the system that will be placed in the control room or B.M.S. room.
12. Details of the earthing system should be included in the submittal.
13. Integration with B.M.S. allows control and monitoring of the smart lighting and dimming system from either the lighting control panel or B.M.S. based on project's specifications.
14. Ensure compatibility with all other related and interfaced systems for communication protocol.
15. Ensure different lighting control and operation for selected or specific areas (service area, staircase, lobbies, corridors … etc.).

Workshop drawings for Smart lighting and Dimming Control System

The consultant should instruct the contractor to prepare the workshop and coordination drawings for the Smart Lighting and Dimming Control System to be submitted for the consultant's approval. The followings should be considered in the drawings:

1. Consider spares for future expansion (Refer to the project specifications).
2. Workshop and coordination drawings for S.L. and D.C. systems should be submitted with lighting drawing clarifying location of lighting circuits, heat and motion detectors, photocells, override switches, staircase lighting, containment for power and control wiring, control wiring and Equipment-related to the control of smart lighting and dimming systems, power supply circuits reference and mounting height for all switches and dimmers ...etc. Coordinated with all other services (Architectural work, finishing, false ceiling, etc.).
3. Riser diagram for S.L. and D.C. system with all details, including all activities connected to the system like power supply circuit reference, connection to B.M.S., F.A.S., Emergency light, lighting circuit ...etc.).
4. Control room, Security room workshop and coordination drawings with all details showing location of Main Control Panel, monitoring panel, power supply circuits' reference, wires type, Dimmers' location and all components in the control room related to the system with dimensions and coordination with other services ...etc.
5. Shop drawings to be signed by the suppliers of the material. The signature gives indication that the supplier specialist is accepting the installation and the installation details are matching the supplied system.
6. Flow chart for all activities related to the system showing all interlocks and interfaces with other systems and scope of all involved contractors.
7. Details for incoming and outgoing wires/ cables for the control panel and dimmers should be shown in the workshop/ coordination drawings. Wiring to be identified by ferrules.
8. Block diagram for S.L. and D.C. system showing all systems details, including all activities connected to the system like connection with F.A. system, emergency lighting, B.M.S., access control, control panel, ... etc.
9. The control panel should be capable of interfacing with access control F.A.S., B.M.S...etc. Ensure compatibility of communication protocols.
10. Number of motion, heat detectors and photocells should be correctly calculated. Successive Detectors' scan should be overriding each other to enable continuous lighting for the walkers without interruption due to dead zones.

11. All Power supplies required for S.L. and D.C. system should be from separate section, ELCB/MCB and from emergency D.B. if available.
12. All override switches should be near the lighting control panel.
13. For the staircase, a timer contactor circuit should be used (based on the project's specifications). The circuit should be able to switch the stairs from above and beneath floors.
14. All wiring for control and power should be according to the standards, project specification and authority regulation. Voltage drops calculations to be considered.
15. The single-line diagram should indicate the location of the lighting point, control panel, all junction boxes inside electrical room, the power supply circuit reference, and interface with access control system B.M.S., F.A.S... etc., with reference to the floor level in the building.

Installation of Smart Lighting and Dimming Control System

The consultant should follow the contractor to ensure the installation of the following activities and devices for the S.L. and D.C. systems with all related and interfaced systems and components, the contractor should check and ensure the followings:

1. The contractor should submit an installation method statement for the whole system with all related interfaces and components.
2. Ensure that all conduits and wires are installed between the lighting point, D.B., control panel, interfaced system, etc.
3. Ensure approval for the philosophy of operation and workshop, coordination drawings by the consultant before starting installations.
4. Ensure that the contractor uses proper back boxes for the installed components.
5. Before the final fix (fixing of components), ensure all received materials are as per approved material submittal and authority requirements.
6. Proper protection for all devices, lighting point, wire, and junction box shall be provided during installation work.
7. Final location and mounting height of lighting point, main panel, override switch, dimmers ... etc. should be according to the approved workshop and coordination drawings or approved mounting height schedule.
8. All circuits, wiring, junction box, control panel and other devices shall be identified by using engraved labels and ferrules.
9. The override switches' location should be installed according to the approved coordination drawings.

10. System earthing should be according to the supplier's/manufacturer's recommendation.
11. A wiring diagram shall be provided inside the control panel with a list of all related equipment and components.
12. The mounting height of devices, switches, dimmers...etc. Should be as per the approved workshop drawing or approved mounting height schedule.
13. Ensure that all terminations are correctly connected and completed to the specified terminals for all devices, control panels, dimmers...etc.
14. Ensure installing control wires linking all ballast. These control wires should be installed in suitable containment segregated from power wiring.
15. Activity chart, work progress and look ahead program of work for S.L. and D.C. System clarifying steps of executing the work (Civil Work, Electrical Work, Finishing Work, testing and commissioning) and defining the scope of work for involved contractors with a monthly update.
16. Segregation between control/ data cabling and power wiring should be maintained for safety and noise immunity.
17. All control wires and power cables shall be with extra length for connection to the terminal equipment.
18. Dimmer Racks should be in a ventilated, dry location. (Consult the manufacturer for dimmer ventilation).
19. Supplier and Contractor's Site Engineer should monitor daily and continuously the installation activity till completion. Periodical progress report showing site status, requirements ...etc. shall be submitted by the contractor.
20. Break down for all items related to the system shall be proposed by contractor for billing purpose; these broken-down percentages should be agreed by all parties.

Testing and Commissioning of Smart Lighting and Dimming Control System.

A. The consultant should instruct the Contractor to complete all pending or delayed work related to the system. Testing and commissioning method Statements to be submitted by the contractor for the consultant's approval including standard readings and the successful test criteria.

B. Complete the wiring megger test before starting any termination for the system components.

C. Confirmation letter from the supplier stating that S.L. and D.C. System installation is completed and ready for operation and testing.

D. Visual inspection and check for the system to ensure that proper installation and termination are completed and to ensure the followings:

1. All components are selected according to the approved material submittal and installed according to the approved workshop/ coordination drawings.
2. Correct and coordinated location of all controller and lighting points…etc.
3. The power supply is connected as per system requirements.
4. The system is connected to F.A.S., B.M.S…etc.
5. The main control panel and all other devices are identified and labelled by using engraved labels and ferrules.
6. Mechanical checking for all components inside the control panel.
7. Ensure installing control wires linking all ballast. These control wires should be installed in suitable containment segregated from power wiring.

E. Function test to ensure:

1. Operation of all dimmers for all channels with 100% loading output.
2. All lights are with full brightness.
3. Addressing and programming each device on the network according to the approved lighting control schedule.

F. Final approval for the system is subject to successful testing and commissioning of the system.

Operation and Maintenance (O&M) manual for S.L. and D.C. lighting.

The consultant should follow the contractor to submit the O&M manuals for the Public Address system, including the following requirements:

1. Description of the project system.
2. How to use the manual.
3. Troubleshooting guide for system breakdown.
4. Safety instruction at work.
5. Start-up and shutdown procedure.
6. An original catalogue highlighting the used material with part number.
7. The as built drawing includes a block diagram and a wiring diagram for the main control panel.
8. List of spare parts.
9. Procedure for routine and preventive maintenance for monthly/1 year/5 years.
10. All guidelines for using devices and controllers inside the control room.

11. Warranty certificate for 2 years based on project specifications.
12. Testing and commissioning certificates/reports.
13. Emergency information.
14. Manufacturer/supplier information and contact details.
15. Overview and mode of operation (sequence of operation).
16. Training for security (Training can be done by submitting a training manual for the consultant and client's approval).
17. Programming, set up, and system configurations with used software.

Note: For Façade Lighting in Towers, nearly similar comments for the smart lighting, the only point is requiring coordination for the following:

1. The riser diagram extends to all floors with all connections and interfaces.
2. Containment from electrical room that contains the panels to the outside lighting that enable wiring and maintenance without entering to the flats. (The containment should be continuous through the façade by coordination with façade supplier and façade installation).
3. Location of the controller to be selected to enable full control (in case wireless control, check the requirement for antennas on the façade and these antennas shouldn't be visible. Also, the signal shouldn't interfere with other services).
4. Coordination with a Cradle cleaning system to be used for facade lighting maintenance.
5. Lighting fittings installation requirements in the façade to be coordinated with façade supplier to provide the requirements (junction box and light base).

Guest Room Management System GRMS

Approval of material submittal for Guest Room Management System

The consultant should follow the contractor in arranging periodical meetings between the electrical engineer and the supplier of the GRMS system. The meeting should discuss and clarify the mode of operation and any requirements for installations. The contractor should be instructed by the consultant to provide the followings to get the consultant's approval for the material submittal:

1. The selected brand or manufacturer is listed in the consultant's approved manufacturer list.
2. A brief description of the system and sequence of operation.
3. Compliance statement to be submitted. The supplier should clearly show his compliance to the specifications, non-complied sections to be highlighted with justifications.
4. The submittal should include the original catalogue highlighting the selected components and the technical points complied to the project specifications. Also, to include the sequence and philosophy of operation.
5. A block diagram for the GRMS system showing all system details and connections to B.M.S. and any other systems according to the project specifications.
6. The submittal should include the wiring diagram for the control panel, details of terminal blocks, and details for incoming and outgoing wires with entry details.
7. Valid supplier trade license with expiry date, with ministry of trade certification as a legal supplier of the material should be included in the submittal.
8. The used cables should be submitted for approval based on supplier's recommendations.
9. The control panel should be capable of being interfaced with other systems like B.M.S. and Access control.
10. The earthing requirement should be shown in the submittal.
11. Communication compatibility with B.M.S. and all interfaced systems.

Workshop Drawings for GRMS

The consultant should instruct the contractor to prepare the workshop and coordination drawings for the Guest Room Management System to be submitted for the consultant's approval. The followings should be considered in the drawings:

1. Spares should be considered in the system for future expansion (Refer to the project's specifications).
2. Workshop drawing to be attached with lighting drawings. Coordination drawings for GRMS should show the locations of lighting circuits related to the system, equipment related to the control with power supply reference, dimmers and switches showing the mounting height of the mentioned components

and coordinated with all other services and furniture layout.

3. Shop drawings should be signed by the material suppliers. The signature gives an indication that the supplier specialist is accepting the approved installation drawings and that the installation details match the supplied system requirements.
4. Wires and cable should be installed according to the standards and the authority regulations and to comply with project's specification taking into consideration the voltage drop within the wires.
5. System earthing should be done according to the supplier's recommendations.
6. Junction box distribution should be coordinated and decided according to the type of the constructed building or room and should be coordinated with furniture layout.
7. The final location of the distribution board should be decided based on the approved coordination drawings, keeping enough space for future expansion.
8. Lighting, shutters, and A.C. controllers should be installed near the main entrance of the room.
9. If touch screen is part of the supplier system, then to be shown in the drawings in suitable place with easy accessable location.

Installation of GRMS

The consultant should follow and instruct the contractor in the installation stage of the guest room management system with all related and interfaced systems to check and ensure the followings:

1. Suppliers should submit method statements for installation activities for the whole system with all related interfaces and components.
2. The contractor should ensure the consultant's approval for the philosophy of operation of the system before installation on the site.
3. Before installation, ensure all materials are as approved in the submittal and without any damages.
4. The contractor should use appropriate back boxes for all components in the installed system.
5. Wiring diagrams should be provided inside the control panel.
6. Proper protection for all wiring and devices shall be provided during installation.
7. If the touch screen is part of the supplier system, then it should be installed in a suitable position to be accessed easily.
8. Activity chart, work progress and look ahead program of work for GRMS clarifying all steps for executing the work such as civil work, electrical work, testing and commissioning, finishing work and defining the scope of work for the involved contractors with monthly update.

9. Engraving labels and ferrules should be used to identify cables and wires for all internal and external components.
10. Supplier, Contractor's Site Engineer should daily and continuously monitor the installation activity till completion. Periodical progress report showing site status, requirements ...etc. Shall be submitted by the contractor.
11. Breakdowns for all items related to the system shall be proposed by the contractor for billing purposes; these breakdown percentages should be agreed upon by all parties.
12. The mounting height and location of the components should be as per the approved coordination drawing or approved height schedule.
13. All termination for all devices and control panels are completed and tightened properly to the correct terminals.
14. Segregation between control data cabling and power wiring should be maintained for safety and noise immunity.
15. All control wires and power cables shall be with extra length for connection to the terminal equipment.

Testing and commissioning for GMRS

A. Before start testing of GMRS system, the consultant should instruct the Contractor to complete all pending or delayed work related to the system. Testing and commissioning method Statement to be submitted by the contractor for the consultant approval including standard readings and successful test criteria.

B. Complete the wiring megger test before starting any termination for the system components.

C. Confirmation letter from the supplier stating that the GMRS system installation is completed and ready for operation and testing.

D. Visual inspection and check for the system to ensure that proper installation and termination are completed and to ensure the followings:

  a) All components (control panel, controllers, light points) were installed according to the approved submittal, workshop, and coordination drawings.
  b) The power supply is connected according to system requirements.
  c) The main control panel and all other devices are identified and labelled using engraved labels and ferrules.

E. Function tests should be carried out for the system to ensure the following:

  a) Check all lights are operating properly.

b) Check all sockets are operating properly.

c) A.C. is working as per the specifications, and all modes of operation are checked.

d) Check the interfacing with B.M.S. and other systems.

e) Final approval for the system is subject to successful testing and commissioning.

Chapter 10

Installations (Part 5) Additional Systems

The following chapter will explain the installation work of additional systems that required electrical power and coordination work with civil and architectural work, these systems are:

1. Garbage chute.
2. Cradle (window cleaning) system.
3. Building Management System.

Garbage Chute System.

Approval of material submittal for Garbage Chute System

The consultant should follow the contractor to arrange periodical meetings between the Electrical engineer, the supplier of the Garbage Chute System and the suppliers for related systems like the Fire Alarm and BMS to discuss and clarify the modes of operation in normal conditions, fire and power supply failure, material submittal work progress etc. The contractor should provide the followings for the consultant's approval for the material submittal:

1. The selected brand or manufacturer is listed in the consultant's approved manufacturer list.
2. A brief description of the Garbage Chute system and operation.
3. Compliance statement to be submitted. The supplier should clearly show his compliance to the specifications, non-complied sections to be highlighted with justifications.
4. Flow chart for all components and devices, including the control panel, conduits, wiring, location, type, power supply … etc. and any other interfaced system with Garbage Chute (Fire Alarm, firefighting, BMS) according to the project specification and authority requirements.
5. The submittal should include the original catalogue highlighting the selected components and the technical points complied to the project specifications. Also, to include the philosophy and mode of operation.
6. Valid supplier trade license with expiry date, and ministry of trade certification as a legal supplier of the material in the submittal.
7. Block diagram for Garbage Chute system showing all details including all activities connected to the system like power supply reference, and all other interfaced systems.
8. Details of the earthing system should be included in the submittal.
9. Troubleshooting procedure in case of power supply shut down to be part of the submittal.
10. Identification with engraved label and ferrules for all cable, accessories, and control wire to be included the submittal.
11. Details of the control panel should be submitted, including the followings:
    a. The cleaning cycle shall be activated through a push-button switch located on the control panel.
    b. An emergency stop push button with manual reset should be provided on the panel.
    c. The control shall include an individual floor monitoring system.

   d. Flushing and compaction timers should be included in the control panel.

   e. The control panel should always show the status of individual floor intake doors (A forgotten open door or improper door closing will give an alarm).

   f. Indicating lamps should be provided for the status of main power supply and control voltage. All indication lamps should be of low voltage type with built in transformer or LED type.

   g. Phase failure and earth leakage protection shall be included in the control panel.

   h. A sub-control panel should be installed in the garbage room to operate the cleaning system, lock the intake doors, and operate the extracting fan.

   i. Provision shall be made to interface the master control panel to the building management system to submit the common fault and alarms of the garbage chute.

   j. All outgoing circuits, including the chute extract fans, shall be controlled through the control panel. Status of the circuit should be shown.

   k. The hooter alarm should be connected to the control panel and fixed inside the guard room or outside the garbage chute room.

   l. The connection of the FAS system should be in 2way:

      i. The signal from FAS to the control panel is a fire in the building or garbage room.

      ii. Signal from the control panel to FAS if a fire exists inside the garbage chute itself.

12. The actual load for the system should be submitted.

13. Block diagram/riser diagram for all activity of the system and interlock with BMS, FAS to be part of the submittal.

14. The control panel enclosure should be weatherproof type.

15. Communication protocol should be compatible with other interfaced systems and BMS.

16. An indication lamp should be provided for each door (green and red).

Workshop Drawings for Garbage Chute.

The consultant should instruct the contractor to prepare the workshop and coordination drawings for the Garbage Chute System to be submitted for the consultant's approval. The followings should be considered in the drawings:

1. Enlarged coordinated workshop drawings (plan and section details) for main/sub-garbage chute room and shaft/ riser clarifying all requirements and all interlocked systems to Garbage Chute including:

   a. All dimensions.

b. Size of conduits and wires (number of wires and wire sizes).

c. Junction box size and details.

d. Location of outlet points with dimensions/ height

e. Electrical schematic diagram and Electrical data.

f. Size of control panel and location

g. Rating of garbage control panel isolator, location, and the source of feeding power supply.

h. The interface between the garbage control panel and other systems like BMS, FAS, etc.

i. Location of Ex-fan isolator. It should be installed with a separate conduit and power supply.

j. Location of FCU/ thermostat, refer to Mechanical drawing.

k. Garbage Chute containment, all conduiting inside riser or garbage room shall be concealed (recessed)PVC conduits. if surface installation is required then it should be G.I. conduits.

l. Location of trolleys.

m. Main power supply details and isolator.

n. Switch/socket outlet/emergency light/normal light, etc., weatherproof type.

2. Block diagram/riser diagram for all activity of the system and interlock with BMS, FAS to be included in the shop drawings.
3. Shop drawings, details, and wiring diagrams should show the number of wires and the required size.
4. Preparation of builder work drawings related to the system in detail, showing all openings with dimensions. Location of all sleeves related to other systems interfaced with the garbage chute.
5. Power supply for the Garbage Chute Panel should be according to supplier specialist requirement, fed from separate section (ELCB and MCB). ELCB should be of 30 mA and the supply shouldn't be shared with other system.

Installation work for Garbage Chute

The consultant should follow and instruct the contractor in the installation stage of the Garbage Chute system to check the followings:

1. The contractor should submit an installation method statement for the system for the consultant's approval.
2. All safety precautions should be followed during installation activities according to the project safety plan.

3. Installation activities should follow the approved method statement and should include the followings:
    a. After the typical floors Structure work completion, the supplier should visit the site to take and record the as-built measurements.
    b. Fabrication of chute trunking and intake throats shall proceed in supplier Workshop. Coating with Revac-Dc-1010, 2mm thick should be applied.
    c. After fabrication completion, the material shall be delivered to the site, and MIR (material inspection request) will be raised for inspection by the consultant before installation.
    d. The installation of firefighting pipes, water cleaning pipes, and electrical cable ducts should be coordinated.
    e. Builders' work details are to be followed based on the Approved Shop Drawings.
    f. For fabricated hooper doors inspection form should be submitted upon delivery of hopper doors to the site.
    g. Install hopper doors and bottom-offset elbow (Velocity Arrestor) in the Garbage Collection Room.
    h. Installation of control panel
    i. System Testing and Commissioning.
4. The main contractor of the system should ensure full protection of the material inside the store.
5. The Main Contractor of the system should make full protection for the system during the Construction. The Garbage Chute shaft should be clean from water and other services.
6. The contractor of the system should provide engraved labels and identification for all junction boxes, isolators, control panels, interfaces with other systems, emergency push buttons, etc.
7. All Electrical installation works in garbage room and the shaft shall be done using G.I. containments and heat resistant wires.
8. Inspections for the following to be done:
    a. Opening dimensions and plumb should be inspected before work commences.
    b. Inspection of delivered material, consultant approval is required on official inspection request.
    c. Inspection of chute trunk and throat after installation, consultant approval is required on official inspection request.
    d. Prior to shaft closing, MEP activities (by others) should be inspected, and clearance should be Issued to proceed.

e. Inspection of the complete system.

f. Any other specific requirement may be required by the client.

9. Activity chart and program of work for the garbage chute system should include the execution steps of the work like Civil work, Electrical work, finishing work, testing & commissioning, and scope of work with monthly updates.
10. Supplier/ Contractor's Site Engineer should monitor daily and continuously the installation activity till completion. Periodical progress report showing site status, requirements ...etc. Shall be submitted by the contractor.
11. Breakdown for all items related to the system shall be proposed by the contractor for billing purposes; these broken-down percentages should be agreed upon by all parties.
12. The contractor must ensure that all safety requirements are in place. Access to the garbage chute room should be clean and dry, complying with the project safety plan.

Testing and commissioning of Garbage Chute System

1) Before starting the testing of the garbage chute system, the consultant should instruct the contractor to submit a specific method statement for testing and commissioning, including standard readings and successful criteria.
2) General:
   - Ensure (physically) all the Intake Garbage Chute doors are properly closed.
   - Start the power supply to the system by switching "ON" the Isolator.
   - Switch "ON" the Control Panel of the Garbage Chute System in the Garbage Collection Room.
   - Switch "ON" the Exhaust Fan from the Control panel; the Fan is installed at the Top of the Roof.
3) Cleaning System (By Sprinkler and Brushing Device):
   - Check physically the availability of water in the tank and disinfectant liquid in the disinfectant tank.
   - Switch "ON" the Chute Cleaning from the Control Panel in the Garbage Collection Room.
   - The Solenoid Valve will be off automatically after 30 seconds.
   - The Cleaning system is activated at the time Solenoid Valve is open, Sanitizing Pump also working, all the doors are locked.
   - In addition, brushing device will start working with sprinkler and it will go down to its own position automatically.

- The Red Pilot lamp will glow when the Solenoid Valve is "ON" Cleaning is in process through the Cleaning Sprinklers installed in every hopper door.

4) Door Locking System:
    - The door will be looked during cleaning and maintenance.
    - All Doors can be locked manually by the Control Panel.
5) One Door System:
    - Only one door can be opened at one time.
    - If one door is open/in use, all other doors get locked automatically.
6) Foot Pedal: The door can be operated by a door handle or by using a foot pedal.
7) Exhaust Fan: The exhaust fan will be operated 24 hours from the control panel with an option of a 24-hour timer.
8) Final approval is subject to the successful commissioning of the system.

Operation and Maintenance Manual for Garbage Chute.

The operation and maintenance manuals should include the followings:

1. Description of the project system.
2. How to use the manual.
3. All guidelines for using devices and controllers.
4. Troubleshooting guide for emergency breakdown.
5. Safety instruction at work.
6. Switching on and off procedure.
7. Original catalogue highlighting the material used with part numbers.
8. As-built drawings, including wiring, riser diagrams and control diagrams.
9. List of spare parts and special tools.
10. Procedure for routine maintenance for monthly/1 year/5 years (as per project specification).
11. Warranty certificate for 2 years (as per project specification).
12. Testing and commissioning certificates/reports.
13. Emergency information.
14. Manufacturer/supplier information and contact details.

15. Configurations, modes and sequence of operation, settings, software, and programming.
16. Training for security, operation, and maintenance teams. ((This can be done by submitting a separate training manual including training details and session content).
17. Description of interfacing with other systems.

Cradle (Window Cleaning System).

Approval of material submittal for Cradle System

The consultant should follow the contractor to arrange periodical meetings between the Electrical engineer, the supplier of the Cradle System and the suppliers for related systems like the Fire Alarm and BMS to discuss and clarify the modes of operation in normal conditions, fire and power supply failure, material submittal work progress in addition to the Authority requirements etc. The contractor should provide the followings for the consultant's approval for the material submittal:

1. The selected brand or manufacturer is listed in the consultant's approved manufacturer list.
2. A brief description of the Cradle system and mode of operation.
3. Compliance statement to be submitted. The supplier should clearly show his compliance to the specifications, non-complied sections to be highlighted with justifications.
4. Flowchart for all activities, components and devices related to electrical work, including the control panel, conduits, wiring, location, power supply ... etc. and any other interfaced systems with the Cradle system (Fire Alarm, BMS) according to the project specification and authority requirements.
5. The submittal should include the original catalogue highlighting the selected components and the technical points complied to the project specifications. Also, to include the philosophy and mode of operation.
6. Valid supplier trade license with expiry date, and ministry of trade certification as a legal supplier of the material should be provided in the submittal.
7. Block diagram for Cradle system showing all details, including all activities connected to the system like power supply reference and all other interfaced systems.
8. Earthing details for the Cradle System should be included in the submittal.
9. Troubleshooting procedures in case of power supply shutdown should be included in the submittal.
10. Identification with engraved labels and ferrules for all cables, accessories, and control wires should be included in the submittal.
11. Emergency landing device to be part of the submittal (sequence of operation during power supply failure).
12. During operation, the system should send an alarm to avoid accidents involving human life or property.
13. Based on the civil aviation authority, the operation of the system should be through wireless remote control.
14. The submittal should show the actual load of the system.
15. The power cable used for the system should be submitted for approval.

16. Power cable support clamps should be provided by cradle supplier. It should be used adjacent to the power socket (type of sockets to be coordinated properly with the system supplier).
17. The safety bar should be available on the window side in two places.
18. The cradle runway should be free from any Electrical/ Mechanical equipment, pipe cable, light fixture, Etc.
19. Details of the control panel should be included in the technical submittal. The control panel should include the following:
    1. Panel ingress protection should be I.P. 67.
    2. An automatic cable realer system should be included for operation vertically and horizontally.
    3. The control panel should have the facility to be operated by wireless remote control based on civil aviation regulations.
    4. During operation, the control panel should generate an alarm to avoid accidents.

Workshop Drawings for Cradle System

1. The contractor should submit Roof coordination drawings showing the followings:
    a. Power supply for all points related to the system with Elevation showing height for all points.
    b. All still supports that fixing and hanging the system from inside or outside the Roof.
    c. All other installations or systems in the roof should be coordinated with the cradle system for proper operation.
2. The power supply required for system operation (400V/ 50Hz) should be from separate breaker/ ELCB section. The supply shouldn't be shared with other system and preferably supplied from emergency panel.
3. The workshop drawing should include the followings:
    a. Location of power supply.
    b. Type of power socket (3 phase or single phase).
    c. The type of cable used between the socket and the control panel.
    d. The location of the foundation and socket mounting with the cable support clamp should be shown in the workshop or coordination drawings.
4. Wiring diagram for control panel including terminal block in detail for all wires (incoming and outgoing) from the control panel.
5. The earthing system should follow the standard and supplier recommendations.
6. The final location of the power supply socket should be submitted (in detail) for approval.

Installation work for Cradle System

1. The contractor should submit an installation method statement for the system for the consultant's approval.
2. Activity chart and program of work for window cleaning System including the steps for executing the works such as Civil, Electrical and Testing & Commissioning work defined scope of work with monthly updates.
3. Supplier/ Contractor's Site Engineer should monitor daily and continuously the installation activity till completion. Periodical progress report showing site status, requirements ...etc. Shall be submitted by the contractor.
4. Break down for all items related to the system shall be proposed by contractor for billing purpose; these broken-down percentages should be agreed by all parties.

Testing and commissioning of Cradle System

1. Testing and commissioning method statement showing successful criteria to be submitted for consultant's approval.
2. Visual check for all installed parts ensuring they are installed correctly and no damges occured due to installation activities in the site.
3. Insulation test for all cables.
4. Operation test: operate and check all modes of operation and the action in case of failure , fire or power supply cut.
5. Final approval is subject to successful testing and commissioning.

O&M Manuals for Cradle System

1. Description of the project system.
2. How to use the manual.
3. Troubleshooting guide (for emergency breakdown).
4. Safety instruction at work.
5. Start-up and shutdown procedure.
6. An original catalogue highlighting the used material.
7. As built drawing including schematic riser diagram/block diagram.
8. List of spare parts.

9. Procedure for routine maintenance for monthly/1 year/5 years.
10. Warranty certificate for 2 years (as per project specifications).
11. Testing and commissioning certificates/reports.
12. Emergency information.
13. Manufacturer/supplier information and contact details.
14. Overview and method of operation (sequence of operation).
15. Training for the security, operation, and maintenance team.
16. All guidelines for using all devices/controllers, etc.
17. System settings, programming, and configurations.

Building Management System (BMS)

Approval of material submittal for BMS:

The consultant should follow the contractor to arrange periodical meetings between the Electrical engineer, the supplier of the Building Management System and the suppliers for all related and interfaced systems in the project like all low current systems, distribution systems, Airconditioning systems, Etc. to discuss and clarify installation requirements, power supply, interfaced systems, input-output list and matrix, material submittal work progress etc. The contractor should provide the followings for the consultant's approval for the material submittal:

1. The selected brand or manufacturer is listed in the consultant's approved manufacturer list.
2. A brief description of the BMS system and philosophy of operation.
3. Flow chart for all activities related to electrical work with all interlocks and interfaces with other electrical, mechanical, and architectural systems according to the project specifications.
4. Details of BMS earthing should be included in the submittal.
5. Identification with engraved labels and ferrules to all cables/ accessories and control wires should be provided with the submitted system.
6. Details of the control panels should be included in the submittal.
7. The contractor should submit the input-output list showing the signal details based on the project specifications and referring to the supplier advise for the missed signals. The contractor should consult the consultant if there is any missed signals to be added and he may submit a variation to include such signals. This list will be the base for the preparation of the shop drawings.
8. Block diagram/riser diagram for all activities of the system, interlocks and interfaces with all other systems to be included in the submittal.
9. A list of all equipment related to BMS inside the control/BMS room.
10. Ensure protocol compatibility between BMS and all other interfaced systems (to be discussed in the coordination and technical meetings).
11. All outdoor panels (DDC) or inside the pump room should be of IP65.

Workshop Drawings for BMS

1. Workshop drawing should include the followings:
    a. Location and details of all power supplies from EDB or D.B.
    b. Location of all 13A.S.S.O inside control room/ BMS room.
    c. Location of all control panels, DDCs, other equipment and the required power supply with reference D.B.

   d. Telephone outlets for BMS data transfer (if required by BMS supplier).
   e. Wiring diagram for control panel including terminal blocks wiring details (incoming and outgoing) for all panels and DDCs.
   f. Details of BMS earthing should be shown in the shop drawings.
   g. All containment related to the system (cable trays, trunking, sleeves, junction boxes, external work ducts, etc.).
2. Coordination drawings for the control/ BMS room showing all equipment with physical dimensions inside the room (main control panel, monitoring panel, all accessories, etc.) coordinated with all other services.
3. The power supply feeders required for BMS operation should be from separate breaker/ ELCB section with 20A D.P. switch with fuse and indication lamp for each control panel.
4. Schematic diagram and riser diagram showing conduit sizing, junction boxes, wire or cable type and size, power supply requirement, telephone system connection, looping between all DDCs, and clarifying all systems connected to BMS.
5. Input/ output matrix showing all connections (outputs) to all interfaced systems and all conncetions (inputs) from all interfaced systems in the project and according to the project specifications.

Installation work for BMS

The consultant should follow and instruct the contractor in the installation stage of the BMS system to check the following:

1. The contractor should submit an installation method statement for the system for the consultant's approval.
2. The contractor should get the consultant's approval for the workshop and coordination drawings related to BMS systems and its control room.
3. Activity chart and program of work for the system including execution steps for Civil work, MEP work, BMS installation finishing work, testing, and commissioning with monthly update.
4. Supplier/ Contractor's Site Engineer should monitor daily and continuously the installation activity till completion. Periodical progress report showing site status, requirements ...etc. Shall be submitted by the contractor.
5. Breakdown for all items related to the system shall be proposed by the contractor for billing purposes; these broken-down percentages should be agreed upon by all parties.
6. The main contractor should ensure full protection for materials inside the store and after installation at the site.

7. The contractor must ensure that all safety requirements are fulfilled according to the project safety plan.
8. No joint in BMS wiring is allowed in any way.
9. All cables and wires are to be numbered properly for easy identification.

Testing and commissioning of BMS

1. Before starting testing of BMS system, the consultant should instruct the contractor to submit specific method statement for testing and commissioning including standard reading and successful criteria.
2. Usually, the installation contractor tests and commissions the BMS.
3. A testing and commissioning report should be submitted for the consultant, including the T&C results, to be compared with the successful criteria of the T&C method statement.

Operation and Maintenance Manual for BMS

1) Description of the project system.
2) How to use the manual.
3) Troubleshooting guide (for emergency breakdown).
4) Safety instruction at work.
5) Start-up and shutdown procedure.
6) An original catalogue highlighting the used material.
7) As built drawing including schematic riser diagram/block diagram.
8) List of spare parts and special tools.
9) Procedure for routine maintenance for monthly/1 year/5 years.
10) Hardware and software copy for the system.
11) Warranty certificate for 2 years (as per project specifications).
12) Testing and commissioning certificates/reports.
13) Emergency information.
14) Manufacturer/supplier information and contact information.
15) Overview and method of operation (sequence of operation).
16) Training for the security, operation, and maintenance team.
17) All guidelines for using all devices/controllers, etc.
18) Configurations, modes and sequence of operation, settings, software, and programming.
19) The final input / output matrix list.

Chapter 11

General Services Activities

The following chapter will explain the requirements for general activities before and through the installation work of the electrical systems, these general activities are:

1) Clearance before Closing False Ceiling
2) Basement/Parking Area
3) Main and Sub Electrical Room
4) Staircase
5) Kitchen and Toilet
6) Reception and Control Room
7) External Work

Clearance before Closing False Ceiling

Before closing the false ceiling, the electrical consultant should inform the contractor to check and verify the following installations and activities. False ceiling clearance can be signed, and false ceiling closure approval can be granted after completion of the following general and specific comments:

General comments:

1. All termination of power outlet should be near the devices (camera, FCU ... etc.) in accessible area.
2. Power supply wires between the PVC box and devices should be routed through the flexible conduit. **See 10-1**

10-1 Final connection of wires or cables to the load or final component using flexible conduit.

3. For surface conduit, both ends should be close by foam (to avoid passage of water drops to the Electrical point or D.B. due to the condensation resulting from temperature difference under and above the false ceiling).
4. It is recommended to provide a catwalk above the false ceiling for double-height false ceilings (void) for maintenance and servicing of all electromechanical works during and after the completion of the project.
5. The contractor should complete the markup drawings for all systems showing installation deviations from the shop and coordinating drawings due to site conditions; the markup drawing will be used as a reference for the as-built drawings without opening the false ceiling again to check the installations.
6. All identifications for trays, trunkings , wiring and cables, and all engravings for equipment and components are already completed.

Specific comments

In addition, the contractor should check the following activities specifically for the installed systems and ensure completion of the requirements as shown below:

Lighting System

1. Lighting points wiring should be completed as follows:
    a) All Covers for PVC/ GI boxes should be completed.
    b) A suitable adaptor between the PVC/ GI box cover and flexible PVC/ GI conduit should be completed.
    c) All wires to the final connection points should be routed through flexible conduits for wiring protection.
    d) All wirings are connected using suitable connector.
2. A preliminary test for all wires Should be done ensuring no damage will be due to daily site work.
3. The height of false ceiling should be marked at site. Enough space should be provided between light fitting and other services above false ceiling like (FCU, cable tray, drop beam … etc.) to allow installing light fitting and control gear.
4. Re-count the number of lighting points above false ceiling, it should be as per the final approved workshop drawings, and load schedule.
5. Light-fitting drivers are installed in a suitable place close to the false ceiling opening to facilitate future replacement.
6. No strip joint is allowed between wires in the junction box above the false ceiling.
7. Ceiling rose can only be used:
    a. If the light fitting is a pendant in areas without false ceilings.
    b. If the distance between the concrete ceiling and the false ceiling is less than 60cm and the false ceiling is the removable type.

Power System

1. Ensure completion of wiring related to power outlets for (FCU, W/H, Ex-Fan, power for low current system … etc.) as shown below:
    a. A preliminary test for all wires and cables is completed to ensure no damage is caused by daily site work.
    b. Completion of wire Connection/ termination related to the equipment above false ceiling

(electrical and mechanical)

c. Ensure that all power points are wired and completed according to the final approved shop drawing and load schedule.

d. Re-count the number of power points related to all systems above false ceiling, it should be as per the final approved workshop drawings, Low current systems requirements and matching the final approved load schedules.

e. Cable tray work, trunking, G.I. conduits and cable/ wire routing are all completed, sharp edges and bends in trays and trunking are covered with rubber to protect the wires and cables insulation from future cut. Grommets are provided for all wiring entries through cut plates.

f. Marking of cable trays, trunking, cable and wiring identifications are all completed.

g. For as built submission, all markup drawings are completed, showing all site deviations from the approved shop drawings due to site conditions.

Fire Alarm and Voice Evacuation Systems

1. Wiring for detector, break glass and speakers for voice evacuation are completed, Wiring installed on ceiling should be fixed by fire rated strips. Or should be routed inside G.I. conduits or trunking, PVC containment is not accepted.
2. All junction boxes for the system are covered by G.I. cover, and wiring inserted into the junction boxes through adapters.
3. Installation of detectors above the false ceiling is completed. Detector addresses are checked and recorded.
4. Preliminary test of detectors/broken glass and voice evacuation speakers wiring is completed.
5. Re-count the number of points related to Fire Alarm system's components above false ceiling, it should be as per the final approved workshop drawings.
6. Strip/ joint in the Fire Alarm wires is not allowed.

CCTV System

1. CCTV wiring (data and control wires) is completed.
2. Preliminary test of wires (data and control) is completed.
3. Power supply points for CCTV camera are completed. All junction boxes are covered with PVC. Wiring will be routed through an adaptor and flexible conduit.

4. The junction boxes above false ceiling related to CCTV should be fixed in inaccessible areas.
5. All identification (engraving) is completed for the junction Boxes above the false ceiling.
6. Strip/ joints in CCTV wiring are not allowed.
7. Re-count the number of points related to the CCTV system above the false ceiling; it should be as per the final approved workshop drawings.

Public Address System

1. Speakers' wiring to be completed. All junction boxes are covered by PVC cover. Wiring are routed through an adaptor and flexible conduit.
2. Preliminary test of wires is completed.
3. Enough space should be provided between speakers and other services above false ceiling like (FCU, cable tray, drop beam … etc.).
4. Strip/ joints in Public Address System wiring are not allowed.

Telephone and Data System

1. Installation is completed for cable trays/ Trunking for (fibre optic cables) including branches to the corresponding (flats/shops) through block work.
2. Preliminary test of the cables (Fiber optic, Cat6 cables).
3. Provision of identification and engraved labelling for cable trays, cables, trunking and wires inside boxes above false ceiling.
4. Strips/ joints between the wires in the junction box above the false ceiling are prohibited.
5. Identification for consolidating point and double sockets should be provided and completed.

Emergency lighting system

1. Installation of cables/ wires for emergency lights is completed, boxes are covered by PVC box cover, adaptor, and flexible conduit with connector.
2. Preliminary testing of cables are completed.
3. Enough space should be provided between emergency light and other services above false ceilings like (FCU, cable tray, drop beam … etc.).
4. Strips/ joints between the wires in the junction box above the false ceiling are prohibited.
5. Ensure that all lighting points above false ceiling should be according to the final approval for (workshop drawing/ load schedule/ schematic diagram).

6. Identification for wiring and lights are completed.

GSM System

1. Installation of cables are completed according to the final approved workshop and coordination drawings.
2. Preliminary test of cables is completed.
3. The engraved label for the cable tray above the false ceiling is completed.
4. Cable identification is completed.

BMS System

1. All BMS wiring is completed.
2. Junction boxes are covered with PVC cover and suitable adaptor is used.
3. The preliminary test for the BMS system is completed.
4. Trays, trunking engraving, and wiring identification are completed.

Smoke control system

1. Conducting for the smoke control system is completed.
2. Wiring for the smoke control system is completed.
3. The preliminary test for the smoke control system is completed.
4. Trunking, engraving, and wiring identification are completed.

General Checks

1. Providing PVC/ GI cover for all unused PVC/ GI junction/ pulling boxes above the false ceiling.
2. Identification for all pulling boxes, cable trays, control panels, BMS components, power supply sockets … etc., to be provided above the false ceiling.
3. All junction and pulling boxes should be accessible.
4. Fireproof material to be inserted for wall openings of cable trays/ trunkings / empty conduit, balance space in all openings, and closing of unused openings.
5. All wires between points and devices should be routed using a flex conduit with a robust connection. For G.I. containment conduits, the flexible conduit should be G.I. with PVC coating ensuring earthing continuity. (Based on authority requirements, may be earthing connection is required for earthing continuity for the flexible G.I. containment system).

6. As built drawings should show all false ceiling openings dedicated for future maintenance. The openings to be approved by Architect.
7. Final test for all circuits wiring (insulation wire) to be tested by Megger before closing the false ceiling.
8. All temporary services and waste above F/C shall be removed before closing F/C, like (wires, cables, conduits, cable tray/trunk, lighting, etc.).
9. For double height (void) over false ceiling, ensure all work related to the catwalk above F/C is completed.
10. If shops having drainpipe or water supply, accessible GRP attic or tray should be provided for maintenance purpose. The attic should be sloped to drain in case of any pipe leakage.

Basement and Parking area

The following activities should be checked and completed by the contractor before and through installation in the car parking area.

A. The following points should be checked and inspected for co-ordination drawing in the Basement and Parking Area:

1. Routing for all A/C ducts (ventilation system), A/C CHW and firefighting pipes.
2. Location of all FCUs, Jet-fans, Control Panels related to A.C. & Exhaust system, Generator exhaust pipe … etc.
3. Routing for all cable trays and trunking for LV, H.V. and low-current cables.
4. Location of all Electrical points LV (lighting/ power), low current system (FAS, EM lighting, CCTV, Gate barrier and parking management. Etc.).
5. Location of lightning protection test points (in case done inside the columns to the basement).
6. Height of installation for all outlets should be clearly mentioned. (Or to be referred to the schedule of mounting heights).
7. A/C equipment, power supply points, cable trays, water supply pipes, etc., shouldn't be installed along the driveway (subject to site condition).
8. Electrical systems such as Bus Bars, cable trays, conduits, etc., should be installed with expansion joints at crossings of the expansion joint of the building.
9. Location of all Electrical devices (Intercom system, access control system, parking management system and music system) should be coordinated with mechanical equipment and other electrical systems.
10. High tension cables should be protected with attic or necessary cover as per authority regulation/ requirement. No services are permitted to pass below H.V. cables.
11. All electrical outlets and accessories should be weatherproof type.
12. The Finished Floor Level (FFL) should be mentioned on the drawing for the Basement Floor and outside the building.
13. CCTV system should be as per the authority (Police) approval.
14. Sound with flusher should be inside all electromechanical service rooms and parking areas.
15. The height of the ceiling in parking or service areas related to the basement, like lift lobbies, to have special consideration regarding Fire alarm and emergency light systems to provide additional devices in high levels to provide lighting in all lift shafts or over false ceilings in service areas.

16. Ensure that all distance between H.T. cables, L.T. cables, bus bars duct etc., as per Authorities regulation. The distance between these services and the room ceiling should be minimum 40cm.

17. The final location of all control panels should be as per authority regulation.

18. The main contractor should conduct a full study of earthing and lightning protection before starting construction work, fully coordinated with other services, sub-contractors and suppliers of the system, and authorities' requirements.

19. The main contractor (MEP coordinator) must arrange a meeting with all sub-contractors to get all types of activities inside basement and to complete coordination between all activities before the start of construction work.

B. The following points should be considered in Builder's work drawings:

1. Up-stand/Threshold should be provided in front of doors for all Electrical and Mechanical service rooms (Electrical rooms, Transformer rooms, RMU rooms, MDF rooms, Telephone rooms, Pump rooms, AHU rooms, Mechanical rooms ... etc.) to prevent water entry into these rooms in case of flooding.

2. Ensure sleeves are provided for all services passing through drop beams as per approved workshop drawings.

3. All sleeves between the Basement wall and outside the building should have a puddle flange if the water level is high. (For Power, External lighting, Telephone system etc.)

4. Extra sleeves between the Basement and outside the building shall be provided for future use.

5. All sleeves between the water tank and pump room should be provided with a puddle flange. **See details of puddle flange installation in 10-2**

10-2 Puddle flange installation for water tanks1

6. All sleeves/ openings should be closed with fireproof material.

C. The following points shall be taken into consideration for Lighting drawings:

1. Universal screws should be used to fix the surface-type light fixtures in the ceiling to reduce the effect of vibration.
2. Location of lighting point should be coordinated with all other services. Installation of light fitting above duct, cable tray, machines, equipment, etc. is not acceptable.
3. All lighting points or other electrical services on columns should be coordinated with mechanical services to avoid overlapping of systems. (Example: Drainpipes, Extract Ducts etc.).
4. All lighting circuits shall be verified for the number of lights in each circuit, type of light fitting, protection in D.B. (MCB) and confirmation of rating/ wattage of light from the supplier, especially for metal halide light fittings.
5. Lighting circuit distribution shall be coordinated for the D.B. Schedule and Lighting Control System schedule.
6. Ensure that lighting circuits related to service area, lift lobbies, staircases and lift shafts are complied to the approved mode of operation, the same shall be considered in lighting control system.
7. Extra conduit for all entrance/staircase/escalator area should be provided for signage (exit sign) power supply provision.
8. Containment for the carbon sensor of the extract fan should be considered.

D. The following points shall be taken into consideration for Coordination with Mechanical systems:

1. An overview of all mechanical systems should be understood by the electrical engineer in coordination with the mechanical engineer prior to starting the workshop drawings to understand the mechanical systems requirements and coordination.
2. All Mechanical systems (sprinkler system, chilled water pipes and drainage pipes) should be below electrical systems such as cable tray, cable trunking, etc. and away from outlets such as lighting fixtures, emergency lights, fire alarm devices, CCTV cameras, busbar ducts, etc.
3. All Air-conditioning/ ventilation ducts shouldn't be close to or above electrical systems.
4. Drain point or water pipes inside any authority room (Electrical room, Telephone room, Generator room etc.) is not permitted. Attic Slab shall be provided in unavoidable circumstance.
5. The location of the control panel and emergency push button/ switch should be according to the approved authority drawings.

6. For Authorities' rooms, workshop drawings should be approved by the Authority before starting work. Special care and attention for ventilation and cable routing, etc.

E. Separate co-ordination drawings shall be submitted in detail for the following Areas:

1. Mechanical Rooms. (Pump Rooms, AHU Rooms, Exhaust Fan Rooms, District cooling entry rooms, etc.)
2. Electrical Rooms. (Main and Sub-electrical room, RMU and Transformers' rooms, HVSG rooms)
3. Telephone Rooms. (MDF room and Sub-telephone rooms)
4. Generator Rooms.
5. BMS Room and Control Room.

Main and Submain Electrical Rooms

The following activities should be checked during and after the construction of the main and sub-main electrical rooms (The contractor should check and ensure the followings):

1. For recessed D.B.'s, 10 cm cavity should be provided in the room wall for D.B. enclosure installation (wall thickness should be at least 20 cm).
2. The finished floor level should be Sloped away from risers' openings to avoid waterfalls to the lower floor.
3. Concrete upstands around risers' openings should be provided.
4. Fireproof material should be used to close all openings between floors and walls inside service rooms and areas.
5. An attic or double wall should be provided according to the Authority's requirement if there is a wet area above or around the room.
6. Close all empty conduits and balance space with foam.
7. The frame of the KWhmeter panel should comply with the Authority's requirement regarding the type and size.
8. The frame of the KWhmeter panel should comply with the authority requirements regarding the height and installation level.
9. Provide a grommet or adapter for all openings in the trunking to protect the PVC wires.
10. Coordination drawing with full details for main and sub-electrical rooms should be submitted for approval, including plane/ sections/ elevations for all walls, clarifying the activities' details as follows:
    a. Openings for A/C, cable tray, cable trunking, bus duct, LV cables, Etc.
    b. All equipment related to the low current system (BMS/ FAS/ CCTV/ IC/ Access door/ emergency lighting ... etc.) should be shown clearly in the coordination drawing.
    c. All switchgear material should be approved by the Authority (MDB/ SMDB/ DB/ CB/ kWh/ IMS connection ... etc.)
    d. All lighting points/ fixtures/ switches/sockets should be shown in the coordination drawing.
11. Flow chart for all activities to show the status of work progress starting from civil (concrete, block ... installation of cable, fixing switchgear in addition to testing and commissioning.

Staircase

The following activities should be checked during and after the installation/ fixing of conduit and outlets in the staircase (The contractor should check and ensure the followings):

1. Detailed coordination drawing should be submitted for approval showing all activities inside staircase like FCU, opening for A/C (grill), lighting point, lighting fixture, emergency light, exit light, switch for light and sleeve inside lintel, Telephone Jack, pressurization fan and BMS box for FCU. ... etc.
2. Wiring diagram for lighting circuit inside staircase should be submitted for approval including the followings: size of conduit, size and number of wires connected to D.B. and other accessories for controllers (timer, contactor, override switch, driver etc.).
3. All back boxes in the concrete should be installed after completing the concrete in locations where the thermocole was provided before casting the core wall.
4. If the staircase area contains a light fitting installed with FCU, then a minimum distance of 30 cm should be considered between them.
5. Ensure that all staircase lighting circuits are fed from emergency D.B. if the generator is available.
6. Ensure that the location of all fire telephone jacks, broken glass and fire alarm speakers comply with Civil Defense requirements.
7. Ensure that two circuits for emergency light are connecting fittings alternatively (alternative point from different central battery unit circuits).
8. Final location of FCU. inside staircase should follow the Authority approval for the shop drawings (should be provided with fireproof enclosure).
9. All openings (grill) related to the pressurization fan should be coordinated with all other activities inside the staircase.
10. A voltage monitoring relay shall be provided inside service D.B. in case the staircase emergency lighting circuit is fed by more than one service D.B.

Kitchen and Toilet

The following activities should be checked during and after the installation of conduit and outlet in the kitchen and toilets (The contractor should be notified to ensure the followings):

1. Complete coordination drawing to be submitted for approval in details for kitchen and toilet with all elevations of walls clarifying the location of all cabinets, size of tile, water pipe, kitchen appliances, sanitary fixture, switch socket outlet door openings, mirrors, ex-fan and kitchen hood, switch light, cooker switch, and outlet, washing machine switch & outlet, dishwasher switch & outlet, etc.
2. Complete coordination drawing to be submitted for approval in details for ceiling and above F/C services clarifying all activities like FCU, water heater, routing of all water pipe and valve, lighting point heat detector, lighting fixtures, ex-fan, kitchen hood, supply and return air grills, detector.
3. If the electric cooker is part of the kitchen appliances, study the actual load and compare it with the design load.
4. Ensure power outlets are provided inside the pantry for offices or retail.
5. The installation of the conduit and fixing of the G.I. box should be done according to the followings:
    a. All electrical installations must be away from water pipes or valves.
    b. All outlets for the dish washer, Washing Machine, Water Heater switch socket outlet, bracket light, etc., must be above and far away from a water pipe, valve, or drainage pipe.
    c. For the Cooker, the contractor should ensure the followings:
        i. The type of G.I. Box to be used for the switch outlet shall be located 20cm away from the cooker side.
        ii. Wire Size should be 10mm2 minimum as per authority regulation.
        iii. A separate circuit/ section/ ELCB is to be used for the cooker supply.
        iv. Coordinate with the supplier for the required socket outlets (one for the oven, one for the plate).
        v. Kitchen hood should be fed from lighting circuit with separate gang in the lighting switch, (location of fused switch outlet should be between cabinet and false ceiling).
        vi. Ensure that the actual load matches the design load.
        vii. The kitchen hood outlet should be coordinated at the site (connection should be with flexible wire – heat resistance).
        viii. The type of Fire Alarm detector depends on the kitchen type. (Heat detector is used in a closed kitchen. Multi-sensor detector is used in open kitchen.)

d. The Shaver socket and hot dryer must be in separate circuits or parts from the power circuit (Connection to the lighting circuit is not accepted), referring to the authority requirement.

e. 13ASSO for the fridge should be at a high level above the fridge in an accessible location (not the backside of the fridge); it should be clear on the load schedule.

f. 2 meters shall be maintained between outlets in case of different phases R, Y, and B, referring to the authority regulation.

g. All outlets should be as far as possible from the sink (refer to Authority regulation).

h. The installation of back boxes for all tile outlets should be done in parallel with the tiles' work.

6. The height of all outlets and switch sockets should be marked on the coordination DRW (elevation) and approved by the architect, engineer, and client.

7. Doorbell Power supply from separate circuit or part from lighting circuit (to be clear in load schedule)

Reception and Control Room

All the following activities should be checked during and after installation/ fixing of conduit, outlet, devices ... etc., in the reception or control room (The contractor should ensure the followings):

1. The contractor must start coordination meetings with all sub-contractors and suppliers related to the services in the reception and control room (lighting system, power system, low current systems, ... etc.) to discuss and prepare a breakdown activity chart for all systems, devices, outlet, civil works, finishing works, furniture, A/C work, electrical work ... etc. The above systems and wirings should be coordinated with the Architect's work regarding the bench location and the arrangement of monitors on the bench.
2. Main contractor should submit coordination workshop drawings for approval from (Arc. Eng./ Client and Electrical Eng.) in details with dimension (plane elevation, section, false ceiling, above false ceiling, the control bench and location of monitors) showing locations of all activities and height of all power supply points. The drawing to be signed by all sub-contractors and suppliers.
3. A flowchart for all systems should be submitted for approval to clarify the power supply requirement for each system.
4. Ensure that final location of power supply points is in an accessible location and near the system according to approved coordination drawing.
5. Ensure that a power supply point is available for each telephone point, data, and low current system (if required according to the approved drawings).
6. Ensure that the location of all emergency lights and lighting fixtures in the ceiling or bracket matches the approved false ceiling layout. The lux level is correct and according to the design.
7. Wiring diagrams for all systems from suppliers showing (conduit size, wire size and type, location of low voltage transformer if applicable or driver, power supply either fused switch with indicator lamp or fuse with indicator lamp, interfacing with other systems and the required modules or relays for interfacing) to be submitted for approval.

External Work

All the following requirements should be checked during and after the implementation of external work activities (The contractor should be notified to check and ensure the followings):

1. The Contractor should submit a coordination drawing for approval before starting any activity outside the plot, building or villa.
2. Coordination drawing for external work should include all the following services taking into consideration recent design approval from the related Authority.
    a. Routing of all external services like (drainage pipe, house connection chamber, water supply pipe, firefighting pipe, power supply cable [H.V., LV], telephone system cable, data system cable [BMS], low current system cable [MATV, CCTV, gate barrier, I/C system … etc.] A.C. system pipe [CHW piping]), streetlight, irrigation system, landscaping… etc.
    b. Final location of the following activities. (External SMATV DB and MDUs/ Control panel/ feeder pillar/ external lighting pool, manholes … etc.) should be clarified. The location should not be in front of any door/ gate / BTU Chamber or any other service.
3. De watering to be done; before any external work activities if required (see the water table report).
4. For duct Installation, the following should be considered:
    a. Before duct installation, ensure the duct size and type comply with the approved drawings from the consultant and Authority. All accessories like bends or couplers should be factory-made.
    b. All underground ducts should be installed above water pipes.
    c. Externally laid ducts and conduits crossing roads should be encased by a concrete layer.
    d. Consider extra pipe/ duct between a main electrical room and outside the plot/ building/ villa for Authority requirements or any other services in the future like mobile generator, external light … etc.
    e. For external road corridor, full coordination between Main and all other contractors should be done like Landscaping contractor, electrical contractor (power cable H.V. and LV), plumbing contractor (drainage and water supply pipe), telephone contractor (cable for telephone or data), low current system contractors (MATV/CCTV), A/C system contractor (CHW piping … etc.
    f. Check all road crossing sleeves are available according to the Authority requirements. Sleeves and ducts to be encased by concrete. 6mm nylon rope should be inserted inside the sleeve/pipe for cable pulling in future.
    g. Prior to placing the duct, the bed should be cleaned from dirt and other construction debris.

   h. Once the ducts are laid, inspected, and cleaned from particles of earth, sand, or gravel, both ends of the duct should be closed with a proper cap.
   i. Finishing floor level for pavement around the building and corridors should be shown on drawings to provide sleeve entering to the building in the correct level.
5. Areas containing electrical components should not be covered by any type of plants/ trees for easy future maintenance. Water pipes should be routed as far as possible from all electrical services.
6. For manholes, the followings should be checked:
   a. All PVC wires, FAS cables, BMS wires and data wires … etc. passing inside manholes should be protected by using suitable conduit/ ducts and sealed from both ends with foam.
   b. All openings around the armored cables passing inside duct to manhole should be sealed from both end of the duct with foam.
   c. The contractor should provide pesticide inside the manhole so that rodents will not damage the wire/cable …
   d. Provide extra sleeves with different sizes for future (close it by foam or cap to avoid insects or reptile entry).
   e. Manholes should be protected against water to prevent any water leakage inside. Hence, affecting the wires/ data wire signals.
7. For External lighting, the followings to be checked:
   a. The mode of operation should be clear and approved by the Client/ Consultant.
   b. Final location of control panels (service DB/ FPL and any connection box) for external light should be away from water and enclosure should be IP65 with canopy and properly ventilated.
   c. Submittal for lighting fixture should show the followings (final location and height, type of paint, colour, shape … etc.), light fittings to be approved by the Architect and client for the shape, colour and colour temp., and the Electrical engineer for technical approval.
   d. In case lighting pole includes cut out/ fuse all internal wires between fuse and fitting should be included and rated 105°c.
   e. In case lighting projector, weatherproof junction box (terminal box) and flexible cable inside G.I. flex conduit should be used.
   f. The supplier should submit the illumination Calculation for approval.
   g. Earthing system to be shown.
   h. Calculation for voltage drop for all circuits.

i. The workshop drawing should show the final location for fittings with dimensions for spacing between lighting fixtures and drivers/routing of cable/control panel location (D.B. service or FPL), circuit reference... etc.

j. The final location of lighting fixtures should be coordinated at the site with respect to landscaping and should be distributed at approximately equal distances as much as possible to show symmetry.

8. For the Earthing system, the following should be considered:

a. Coordinate with civil for Soil resistivity report to decide the type and the procedure for earthing system connection for (substation, feeder pillars, kWh meter, external light, gatehouse, MATV, pump room, swimming pool, lake, water feature and CCTV).

b. Connect Earth copper cable (ECC) to foundation steel for lightning protection. (Based on project specifications).

c. Ensure the earthing system electrode bar is reaching the water level.

d. Ensure normal earthing results based on the soil resistivity report result.

e. The distance between electrode bars for the earthing system should not be less than 6m.

f. The flow chart should be prepared by the contractor for all types of earthing LV, H.V., low current (MATV, BMS) ... etc.

9. For External activity (general), the following should be considered:

a. All external temporary cables should be installed in ducts and to be removed after completion of the project.

b. All external activities/ control panel/D.B./ disconnecting switch/ feeder pillar ... etc. to be I.P. 65 with canopy, with suitable anti-rust painting inside and outside.

c. All conduits should be connected to junction box in external areas from the top or side to avoid water dropping to the conduit. PVC junction boxes should be recessed in wall as PVC can't withstand the Ultraviolet ray, G.I. boxes can be installed surface mounted, but they will be rusty and damage by time.

d. For landscaping area dedicated electrical load for landscaping equipment, check that the load is covered in the schematic diagram and load schedules.

e. Protection and labels for underground cables should be arranged (cable banks/ warning tapes/ PVC tile/ concrete cover and all other necessary protection requirements).

f. Confirmation letter from all sub-contractors or suppliers clearing that all underground ducts / sleeves are provided as required before covering or closing any area.

g. Extra pipes/ conduits between the building and outside are to be provided for future use.

Chapter 12

Conclusion

The consultant role in the project is a key reason for project success, as he will motivate the project and the contractor based on his experience in a way that leads to:

1. Increased quality.
2. Reduced cost and rework.
3. Accurate coordination with other systems.
4. Smooth progress.
5. Easy future maintenance.

All these achievements can be done by correct directions and by putting into consideration that the consultant is working side by side supporting the contractor to ensure work progress with best quality and shortest duration.

As a conclusion for the consultant role: Although many of the points below are the contractor's responsibility, it is a good practice if the consultant can share his experience and directions for the followings to be considered by the contractor in site activities:

Consultant's general activities

1. The consultant is a key parameter for project success if he works supporting the contractor to achieve the work, reduce difficulties, ensuring smooth progress without sacrificing the project quality, the scope of work and ensuring all parties' and stakeholders' satisfaction.
2. The consultant should advise the contractor's engineer, giving directions, providing solutions based on his previous experience, and warning him of consequences due to incorrect or incomplete work or decisions based on his experience in previous projects (especially for young engineers and engineers with limited experience).
3. The consultant should be available to work with the contractor to solve discrepancies and difficulties through his experience or by consulting the designer to find alternative solutions that meet the applicable standards and authority regulations and meet client satisfaction.
4. The consultant should know that working to save himself from blame as a priority and putting all

problems on the contractor is not a solution to eliminate him from responsibility and blame him for the wrong work, project delays, and problems as such consequences will be all parities problems, not only the contractor.

5. The consultant should have enough idea about the project specifications and contractor's scope of work so he will be always confident of himself when asking for a certain action or work and he can prove that it is in the contractors' scope.

6. The consultant should treat both the client and the contractor with justice and fairness for the variations, whether they are negative variations in the client's favour or positive variations in the contractor's favour. Ignoring positive variations for the contractor will cause the contractor to be careless and affect the work quality, especially if this variation results from a mistake or a wrong direction from the consultant that he wants to cover.

7. The consultant should consider the contractor and project duration and the consequences results for delays in approvals of documentation, or delays in inspections or approvals in general. He should do his best to help the contractor by speeding these requests so as not to be the reason behind the project delay.

8. The consultant should check in the submittals that the submitted material can be installed in the site and the cable entry/ exit and panel access are matching site conditions. The size of the material can be entered into the installation location. He should visit the factory with the contractor to check the material preparations (especially the LV distribution system and panels) to ensure it is built and provided matching the site requirements.

9. The consultant should agree on a suitable system for payment by breaking the installations into several agreed segments with the contractor (For example, conducting 30%, wiring 30% and fixing of material and termination 40% of the payment) so the contractor can submit his payments based on the above breakdown to receive monthly payments. Payment is a very sensitive issue on the site as it is the key requirement for project progress, as it is used to procure the required material and project expenses. (Note: usually, the payment given to the contractor is 80% of the actual work done, and always, the contractor should have money left to force him to continue working and not to leave the work suddenly if he is late and receiving all his deserved money).

Technical Meetings for site construction

10. The consultant should attend the meetings held between the contractor and the suppliers and participate in the discussions for the required specifications and site requirements to put the supplier in a clear idea from the beginning about the site conditions and to reduce the time for approval.

11. The consultant should attend the coordination meetings to ensure correct coordination with all other systems.

12. The consultant should attend the meeting with the client and should transfer the contractor requirements for prior approval that will be followed by official request for official approval.

13. The consultant should meet with his staff (inspectors) to guide them about following the site, the main points to be checked, and the area requiring concentration. Also, he should be their mentor to increase their experience. It is very important to have a strong and cooperative team of inspectors. The relationship with the inspectors should always be good and healthy as the inspectors will transfer all site activities status to the senior. Bad relations may cause the inspectors to work carelessly. Inspectors should be informed that they should inform the senior of everything and not hide any point, even if they made a mistake, made a wrong decision, or acted in the wrong direction. It is always the contractor's responsibility to hand over the project in the final stage with the correct work, and the inspectors will not bear any responsibility for their wrong decision as the contractor should hand over the project according to the correct scope and function based on the approved documents that are complying to the standards, regulations, and approved norms.

Material Submittals

The consultant should ensure the followings in the material submittals:

14. The consultant should ensure the selected brand is listed on the approved manufacturer list. Many consultants have approved manufacturer lists, this list is created when the suppliers submit the following to be approved in the designer vendor list:

    a. The supplier prequalification.

    b. Supplier Scope of Work.

    c. Company size.

    d. Company history of work.

    e. Previous projects.

    f. Financial abilities.

    g. Company size and number of staff and offices.

    The supplier usually offers the consulting company to visit their factory and office to check their abilities and their approvals or certificates to be sure of their power in work, by this the supplier will be approved and added to the consultant approved list, this will eliminate new suppliers without enough experience, low-quality products or one season suppliers from being in the project.

15. The consultant should ensure that the submitted material is complied to the required specification and philosophy of operation in the project specification by requesting the compliance statement. (The compliance statement is a copy of the signed specification printed and marked on each paragraph by

the supplier's handwriting as complied or not complied. The supplier should show the deviations if the item does not comply with the project specifications or project requirements. The consultant will study the deviations and will decide whether they can be accepted, if they need the client's approval, or if they should be rejected.

16. The consultant should ask for the original catalogues for the material and all its components and accessories; the catalogue should show the part number of the used material highlighted so the consultant can verify whether the used materials comply (or not) with the standard, project conditions and whether the contractor is using the accessory in the correct way.
17. The consultant should sit in a coordination meeting with the supplier, and all systems' suppliers interfaced with the system under discussion to check the interfacing requirements and the communication protocol with other suppliers' requirements.
18. The consultant should check the submitted samples and to follow the contractor to check the samples whether it fit to be installed in the site, or not. Mockups should be done for approvals if required.
19. The consultant should ask for the trade license of the supplier, office location, the name of the technical staff for future support and spares. The trade license will prove that the supplier is permanently available for support as the official representative to the manufacture, and he will provide the required spare parts and technical support in future, and the supplier is not of the one season suppliers' type.
20. The consultant should check the cause-and-effect matrix, or the input output list related to the submittal (if applicable) which depend on the interfaced systems and philosophy of operation, monitoring and control in project specification.
21. The consultant should ask for all technical calculations from suppliers, and he should be able to verify them.
22. The submittal should include the test certificates or laboratory technical certification for the material.
23. The consultant should ensure the submitted material is matching the authority requirements and regulations.
24. The consultant should ensure that the submitted material is based on the latest approved shop drawings with all interfaces systems. If any change occurs to the systems, then the submittal should be updated accordingly.
25. The consultant should ensure that the submitted material or components are of the latest developed technology that will be available in the market for many years before being replaced.

Workshop, coordination, and builders work drawings.

26. The shop drawings should be provided using the approved material with its actual dimensions, the

shop drawings should be signed by the supplier specialist to confirm that the installation requirement is correct and accepted, and to be signed by the interfaced systems' suppliers to confirm that the requirements for the interfaced systems are provided.

27. The shop drawing should contain the installation details to the level of the required information by the site foreman.
28. Coordination drawing should show all systems in the area or the room with all dimensions and details, also it should show the containment and the accessories.
29. Builder's work drawings should show the required openings for all electrical services based on the sizes shown in the approved electrical shop drawings.
30. Builder's work drawings should be divided to show openings in the slab for vertical installations between floors and openings in walls for horizontal installations.

About the Authors

Nedal Shawkat Nabulsi

BS.c in Electrical Engineering college

University of Baghdad -1981

Worked in the following companies:

1. International Electric company.
2. Al Anwar Electrical Company.
3. Juma Al Majid Company.
4. Central Bank-Sharjah Branch.
5. Khatib and Alami company.

Hani AlKhayyat

BS.c in Electrical Engineering college

University of Mosul -1985

haniiraq@hotmail.com

Tel & WhatsApp: 00971507997476

https://www.linkedin.com/in/hani-mohammed-saeed-8332922b/

Worked in the following companies:

1. Fujairah National Construction and Transportation (FNCT).
2. EMI Electromechanics.
3. Thermo Electromechanics.
4. Hill International.
5. Parsons.
6. WSP Middle East.
7. AECOM.
8. SERCO
9. Keolis.
10. Etihad Rail.

www.ingramcontent.com/pod-product-compliance
Lightning Source LLC
Chambersburg PA
CBHW042039160726
48282CB00039B/1268
9789360067366